Ethical Research Approaches to Indigenous Knowledge Education

Ntokozo Mthembu
University of South Africa, South Africa

IGI Global
PUBLISHER of TIMELY KNOWLEDGE

A volume in the Advances in Educational Technologies and Instructional Design (AETID) Book Series

Published in the United States of America by
IGI Global
Information Science Reference (an imprint of IGI Global)
701 E. Chocolate Avenue
Hershey PA, USA 17033
Tel: 717-533-8845
Fax: 717-533-8661
E-mail: cust@igi-global.com
Web site: http://www.igi-global.com

Library of Congress Cataloging-in-Publication Data

Names: Mthembu, Ntokozo, editor.
Title: Ethical research approaches to indigenous knowledge education / Ntokozo Mthembu, editor.
Description: Hershey, PA : Information Science Reference, [2020] | Includes bibliographical references and index. | Summary: “”This book explores ethical and moral approaches to conducting research in indigenous knowledge education. It also examines learning trends with indigenous knowledge education”--Provided by publisher”-- Provided by publisher.
Identifiers: LCCN 2019026564 (print) | LCCN 2019026565 (ebook) | ISBN 9781799812494 (hardcover) | ISBN 9781799812500 (paperback) | ISBN 9781799812517 (ebook)
Subjects: LCSH: Ethnoscience--Research--Moral and ethical aspects..
Classification: LCC GN476 .E85 2020 (print) | LCC GN476 (ebook) | DDC 500.89--dc23
LC record available at https://lccn.loc.gov/2019026564
LC ebook record available at https://lccn.loc.gov/2019026565

This book is published in the IGI Global book series Advances in Educational Technologies and Instructional Design (AETID) (ISSN: 2326-8905; eISSN: 2326-8913)

British Cataloguing in Publication Data
A Cataloguing in Publication record for this book is available from the British Library.

Advances in Educational Technologies and Instructional Design (AETID) Book Series

ISSN:2326-8905
EISSN:2326-8913

Editor-in-Chief: Lawrence A. Tomei Robert Morris University, USA

MISSION

Education has undergone, and continues to undergo, immense changes in the way it is enacted and distributed to both child and adult learners. In modern education, the traditional classroom learning experience has evolved to include technological resources and to provide online classroom opportunities to students of all ages regardless of their geographical locations. From distance education, Massive-Open-Online-Courses (MOOCs), and electronic tablets in the classroom, technology is now an integral part of learning and is also affecting the way educators communicate information to students.

The **Advances in Educational Technologies & Instructional Design (AETID) Book Series** explores new research and theories for facilitating learning and improving educational performance utilizing technological processes and resources. The series examines technologies that can be integrated into K-12 classrooms to improve skills and learning abilities in all subjects including STEM education and language learning. Additionally, it studies the emergence of fully online classrooms for young and adult learners alike, and the communication and accountability challenges that can arise. Trending topics that are covered include adaptive learning, game-based learning, virtual school environments, and social media effects. School administrators, educators, academicians, researchers, and students will find this series to be an excellent resource for the effective design and implementation of learning technologies in their classes.

COVERAGE

- Higher Education Technologies
- Hybrid Learning
- E-Learning
- Classroom Response Systems
- K-12 Educational Technologies
- Digital Divide in Education
- Web 2.0 and Education
- Instructional Design
- Online Media in Classrooms
- Virtual School Environments

IGI Global is currently accepting manuscripts for publication within this series. To submit a proposal for a volume in this series, please contact our Acquisition Editors at Acquisitions@igi-global.com or visit: http://www.igi-global.com/publish/.

The Advances in Educational Technologies and Instructional Design (AETID) Book Series (ISSN 2326-8905) is published by IGI Global, 701 E. Chocolate Avenue, Hershey, PA 17033-1240, USA, www.igi-global.com. This series is composed of titles available for purchase individually; each title is edited to be contextually exclusive from any other title within the series. For pricing and ordering information please visit http://www.igi-global.com/book-series/advances-educational-technologies-instructional-design/73678. Postmaster: Send all address changes to above address.

Titles in this Series

For a list of additional titles in this series, please visit:
http://www.igi-global.com/book-series/advances-educational-technologies-instructional-design/73678

Digital Learning Architectures of Participation
Nigel Ecclesfield (FRSA, UK) and Fred Garnett (FRSA, UK)
Information Science Reference • © 2021 • 289pp • H/C (ISBN: 9781799843337) • US $165.00

Participatory Pedagogy Emerging Research and Opportunities
Martha Ann Davis McGaw (McGaw Davis Foundation, USA) and Simone McGaw-Evans (Georgia State University, USA)
Information Science Reference • © 2021 • 150pp • H/C (ISBN: 9781522589648) • US $155.00

Visual Approaches to Instructional Design, Development, and Deployment
Shalin Hai-Jew (Kansas State University, USA)
Information Science Reference • © 2021 • 228pp • H/C (ISBN: 9781799839460) • US $185.00

Teaching Language and Literature On and Off-Canon
José Manuel Correoso-Rodenas (Universidad Complutense de Madrid, Spain)
Information Science Reference • © 2020 • 370pp • H/C (ISBN: 9781799833796) • US $195.00

Handbook of Research on Advancing Language Equity Practices With Immigrant Communities
Lourdes Cardozo-Gaibisso (Universidad ORT Uruguay, Uruguay) and Max Vazquez Dominguez (University of North Georgia, USA)
Information Science Reference • © 2020 • 433pp • H/C (ISBN: 9781799834489) • US $245.00

For an entire list of titles in this series, please visit:
http://www.igi-global.com/book-series/advances-educational-technologies-instructional-design/73678

IGI Global
PUBLISHER of TIMELY KNOWLEDGE

701 East Chocolate Avenue, Hershey, PA 17033, USA
Tel: 717-533-8845 x100 • Fax: 717-533-8661
E-Mail: cust@igi-global.com • www.igi-global.com

Table of Contents

Detailed Table of Contents

Section 1
Novelty of Indigenous Knowledge

South Africa's recent higher education protests around fees and decolonizing institutions have shone a spotlight on important issues and have inspired global discussion. We witnessed similar resistance during apartheid, where African languages and ideas were limited. The educational space was the most affected by clashes between languages and ideas; we saw this in the prioritizing of English and Afrikaans over indigenous African languages and the prioritizing of Western medicine, literature, arts, culture, and science over African ones. This chapter will show how formal education and knowledge production in South Africa has been used as a tool to repress Black people, while discrediting their knowledge systems. This discussion will draw from impepho, which is rejected by Christians because its main use is for communicating with ancestors. The herb has many other medicinal uses, but it is still rejected. African practices are used and revitalized by AIC like the Shembe Church and revolutionary movements like FMF.

The chapter examines possible use of instruments and processes such as ethical clearance in the institutions of higher learning as subtle means of perpetuating inequality and racial prejudice towards the indigenous people of South Africa who had recently emerged from the scourge of apartheid with a hope of democracy ultimately providing not only freedom of association and speech, but also intellectual freedom. Freedom to produce African-based knowledge by Black African intellectuals pursuing their postgraduate studies and academics whose careers are at formative stages. However, their vision of becoming producers of African Indigenous knowledge is thwarted by subtle and invisible activities that are aimed at perpetuate coloniality in the higher institutions of learning. Sadly, ethical clearance process has possibly been utilized to derail research outputs that some of the old guard from historically white universities are uncomfortable to witness, thus continuing to maintain the colonial status quo.

This chapter seeks to map how indigenous people and their indigenous knowledge systems are the most researched and written about in the world, yet they are the least understood. The curriculum of the empire and its scientific explanation justified how indigenous knowledge systems should be approached and viewed as well as who has the authority to justify; hence, indigenous knowledge systems were justified as inferior and not worthy of the standard of European knowledge system. In this chapter, Frantz Fanon's thought will be deployed to illustrate how this division of knowledge justifies the perpetuating dehumanisation of indigenous people under the mask of modernisation and globalisation. By deploying decoloniality, Afrocentricity, and Fanonian thought, this chapter seeks to challenge this curriculum that is based on the history of the conquest of Africa that positioned Africa only as a cradle of slaves and the black bodied as created by God only for the benefit of the Europeans.

This chapter discovers the limitations presented by narrow cultural and moral settings and the possibility of incorporating an indigenous African knowledge systems' (IAKS) ethos to redress past injustices, especially diverse cultural values experienced in countries in the 'global south'. However, the emergence of related protests in communities and student structures in education circles, such as calls to decolonize the curriculum and the #FeesMustFall movement. The effects of colonialism continue to be reflected in social structural settings that uphold those Aristotelian parameters that are notorious for marginalizing the knowledge of the 'other', specifically in the 'global south'.

This chapter interrogates the current teaching and learning strategies in South African higher institutions of learning in relation to the negligence of cultural aspects. For many years now, the general population in South Africa agitated for the amendment or annulment of the current education system as a means of restoring the African norms and values. The suspension and/or expulsion of some African students from schools due to their adherence to African values is a vivid example of how the education system undermines African cultures and traditions. The chapter deliberated on story telling as one of the methods that is relevant for Indigenous Knowledge Systems. On the central argument of the chapter the authors conversed on colonialisation and its impact on education in South Africa. OBE and the revised national curriculum statement and the crises in South African education system are discussed. The chapter ends by interrogating the possibility of using African Languages in the school curriculum.

Preface

For the past two decades in South Africa and other parts of the world, transformation agenda in almost all social spheres including education has been a priority in the development programmes. First has been increase demands for transformation of previous structures that served a narrow cultural viewpoint, not only for key operational purposes but also for gaining strategic advantage. Second, supported by information and communication technologies -ICT, the whole business model for many organizations has been transformed. Whereas in the past institutions including education sector could rely on confining themselves to a singular knowledge system to define their scholarship. Today institutions are increasingly becoming more open and are finding themselves to be strategically disadvantaged if they are confined to a unsocial knowledge base, a narrow cultural perspective. The consequence of advances in information technologies and the changing boundaries of the learning sphere have brought the significance the notions of ethical codes and information to the fore. This is because it is information that helps institutions to realize their objectives and helps in the effective management. In the education model of the past, learning and tuition to a large extend was confined to a particular social group and it was relatively easy to protect it from falling in the hands of those who have not agreed to its lucidity (Bangura, 2012). Because education was utilised as a political tool to advance a particular social programme, it was also possible to ensure, with a relative degree of certainty, that it's content and form and ensure that it was readily accessible to its personnel. In fact maintaining confidentiality, despotism and content were the main tenants for managing security. Today because the nature of the society and scope of transformation process has evolved, education system is not just restricted to maintaining communality, confidentiality, content and humanity. Perhaps as Mukuka, (2010; p 43) point out, cultural input will assist in reviewing a new direction for indigenous and ethicality in the academia.

Changing structures, advances in information and communication technologies and the greater reliance of institutions on information indeed poses a number of possibilities for advancing higher level of scholarship practices. Various attempts have been made to explore the decolonisation of operational research of which would require, among other things, an inclusive curriculum by replacing existing material with examples from a particular milieu. In recent years clearly, organizations have developing adequate policies to deal with the need to incorporate various knowledge systems in the learning environment. For instance, article 27.2 (p) of the Universal declaration of Human Rights by the United Nations: "Everyone has the right to the protection of the moral and material interests resulting from any scientific, literary or artistic production of which he is the author" (United Nations, N.d.). This is evidenced not only by increases in policies promulgated related to the advancement and safeguard of indigenous knowledge systems. For example, in South Africa, Indigenous Knowledge Act 6 of 2019, but also the steady increase of authorities and academics to establish adequate basis for implementation of such policies but also debates on the authenticity of research pertaining to ethics of various knowledge systems. One example is the case of Madiba (2014) where there were 'difficulties' to establish whether translinguaging without consideration of the social orientation by study of Madiba illustrate the extent of indigenous knowledge systems moral standards (Mona & Hurst, 2017). Advances in the research surrounding the incorporation of indigenous knowledge systems in to learning environment has also brought about the need to consider adoption of related ethical code to enable to simplify its classification to avert possible misperception.

In view of the constraints that relates to the introduction of ICT in the education sphere in South Africa, this submission attempts to reveal some of the limitations that inhibit equal access and transformation as well as possible teaching and learning approaches that could be adopted to bridge this quagmire in a changing society (Mthembu, 2018a, p.198). Challenges include the fact that etymological shortcomings of information technology in the realisation of the aspired change in the education system cannot be separated from the prevailing neoliberal societal dynamics, especially the augmented reality (Şimşita et al, 2014, p. 931). Based on the South African perspective it is suggested that it should be used as a guiding principle in formulating a transformative curriculum.

This perspective is viewed as the appropriate approach on teaching and learning in the South African context to enhance learners' experience and to satisfy academic and societal aspirations. However, some constraints relate to affordability and structural inequalities that hamper access to education and the successful realisation of a transformed education system in the country. Both provide some insights into the possibilities of a pluralist/multicultural approach that is likely to be the backbone of the education system in the forthcoming years (D'Aiglepierre, Aubert & Loiret, 2017). Specifically, the blended learning approach, curriculum development and use of ICTs depend on an authentic learning environment in order to nurture and guarantee basic, technical and practical skills.

In guaranteeing the success of the use of ICTs in the learning sphere, blended learning education offers various approaches to ensure improvement of knowledge for learners that provides for academic and societal needs satisfaction. Guidelines and policies must be developed to function as a guide for educators in their endeavours to deliberate on curriculum transformation. Lastly, the successful corporation of ICTs in a transformed education system will be possible when education takes place in a society where there is equality and respect for other people's cultural values.

THE CHALLENGES

It goes without saying that, the narrow cultural approach in the curriculum enabled marginalisation of other knowledge systems of which contribute in the inequality in the learning sphere. And any attempt to deal with the problem demands an adequate understanding of the challenges that exist in the new millennium. Such challenges can be classified into four categories:

- The challenge of establishing a multicultural practice in a diverse society and yet being able to control administrative operations.
- The challenge of establishing a relevant curriculum and procedures that adequately reflect the organizational context and new scholarship processes.
- The challenge of establishing correct structures of responsibility, given the complex structuring of education and relevant assessment activities.
- The challenge of establishing appropriate ethical codes for diverse knowledges that are vested with multicultural education aspects to enable the realisation of the envisaged transformed education system.

Numerous studies (for a summary and review see Tshifhumulo et al's chapter in this book, besides Ramdass (2015); Do Vale, (2016) and Higher Education South African, (2014) have indicated that there is a problem in academic sphere especially with respect to inclusion of the previously marginalised communities' knowledge systems in the curriculum. Research has also shown that many a times that inclusion of the indigenous knowledge values will subvert existing controls of narrow cultural ideals that gain undue advantage essentially because either an opportunity exists to do so or they relegate indigenous knowledge systems (Phaahla, 2014). The problem gets compounded even further when an organization fails to formulate relevant policy to monitor transformation agenda and it becomes difficult to institute the necessary formal changes. This was evidenced in the case of Ntokozo Mthembu in his attempts to introduce indigenous African ethical standards using indigenous languages as the teaching tool in the South African academia. This approach to transformation process tended to encounter resistance and the demand for translation of indigenous related subject to English of which attempt to conceal the viability of indigenous languages as the teaching tool in the learning sphere, internal and external scholars, and regulatory bodies in South Africa (Owusu-Ansah & Mji, 2013). Mthembu's case is illustrative of breaches of control, trust, confidence, and deviations from envisage curriculum change and related African centred research ethics and methods or expectations.

The management of transformation in academic institutions in some institution in South Africa clearly lack transformation policy and change of curriculum of which remain something to be achieved. However, there is nothing new in this scenario, since it has long been established that limitations in applying the transformed curriculum and external controls is perhaps the primary reason for breaches in realising this dream (see Education White Paper 3 of 1997). Failure of management to curtail such limitations, which instil fear and despair to some academics of which create an environment conducive for sustainance of status quo, lack of independent monitoring and control of risk, communication breakdown between managers and teaching staff, and the use of information technology to exclude African languages and expectation to overcome basic communication problems in organizations were other reason that created an opportunity for the demand of translation by the previously advantaged staff members to demand the use of English and Afrikaans as only languages in this instance as 'worthy' for utilisation in the learning environment (Gumede & Biyane, 2016; p69).

There is also the challenge of establishing appropriate ethics policies and procedures that adequately reflect the organizational context and encompass the 'new' African ethical codes processes. Such challenges exist at two levels. First at an internal organizational level where the stifle is increasingly becoming difficult for the develop and implement appropriate ethics policy. Second at a broad contextual level, where it is becoming difficult to rely on traditional legal policies to regulate the establishment of indigenous centred behaviour. At an internal organizational level, there is a problem with respect to establishing indigenous related policies. For one there is a lack of knowledge on indigenous ethical codes within organizations that such a limitation exists. Based on Cultural Integration Project that was established by a university in South Africa in 2008, a research that focused on the challenges, barriers and opportunities in respect of the cultural integration in the curriculum, Ramdass (2015); Iheriohanma & Oguoma (2010) contend that development structures and policies that are championed by the colonialists tend to deny space for the emergent leaders in Africa to espouse their related moral standards, but this approach tends to reflect a parochial and limited scope when it comes to improve the livelihoods of the population of the previously colonised nations. This results in a situation that becomes a breeding space for corruption, marginalisation and irresponsible development, as relevant structures and personal responses seems to address issues in a rather superficial manner. The importance for developing an inclusive policy is also identified by Moyo & Gonye in their contribution for this book.

At a broad contextual level, although a wide range of literature has been written in recent years, however, their nature and scope seem to be at odds with the reality. This scenario becomes clearly visible especially when it comes to the scope of the current ethical codes and its application as they reflect the very contested narrow cultural perspective. There are perhaps a number of other alternatives that can be adopted to correct this one-sided approach as the strategy to widen up scope of scholarship in the higher institutions (Msila, 2007; p 148). In many cases the advocacy of indigenous knowledge values system is not the intellectual challenge of tearing apart knowledge systems but create a space of varied knowledges in the learning sphere.

It is indeed a challenge when dealing with an introduction of other knowledges that were previous marginalised in the space that was designed to cater a knowledge that promotes a particular social group. Many a times disasters occur because of renegades of some staff members especially among the previously advantaged lot. According to the observable experience regarding the rate of transformation in terms of introduction of African related subjects and languages in various institutions in the South Africa, shows that there is no political will from various stakeholders including government (Mthembu, 2018a; p 202).

As a consequence, academic institutions are left on their own to determine the rate and timeframe of implementing the aspired curriculum change. Apparently, this lazes affair approach tends to empower the already advantaged former beneficiaries of the past administration who happens to be in upper echelons and take decisions that favours them. According to a new demographic profile of the country's 25 universities, it reveals that at present, some 49 per cent of academics in South Africa are white compared with the 35 per cent of academics who are black, while another 10 per cent are of Indian and 6 percent mix – race descent (Grove, 2017). However, it is estimated that in the near future the demographics might change this scenario to favour the previously marginalised populace.

SEARCHING FOR A SOLUTION

Solutions to the problem of recognition and application of indigenous knowledge oriented ethical codes in the transformation agenda and education sphere in general. Despite various scholars have placed calls for such orientation, in practice over formalized, a contextual and ahistorical solutions designed in a reactive manner, are still the order of the day. Many a times such ill-conceived solutions mark the prevalence of coloniality, meaning the resurfacing of colonialist ideals that are notorious of undermining collectivism.

Establishing formalized rules is one step that could lead towards a solution for managing transformation in the education sector and society in general. Such formalized rules may take the form of security policies that help in facilitating bureaucratic functions in order to resolve ambiguities and misunderstandings within organizations. Both academics and practitioners have made numerous calls for formulating security policies and many of these calls have stopped at just that. Although security policies are essential for laying down rules of conduct, success of transformation programme is clearly a function of the level of their integration with the strategic vision. If we accept that a secure environment is an enabling condition for the smooth running of an innovativeness, then security considerations are a strategic issue and there is a need to configure them for maintaining the consistency and coherence of organizational operations.

In the past transformation policies have been formulated based on checklists and hence tend to identify specific responses to specific conditions. However, if institutions want transformation of curriculum management to be an antecedent to a highly integral education environment, focus needs to shift towards creating a scholarship vision and strategy where adequate consideration is given to the threats and weaknesses of the infrastructure within the broader scope of learning sphere. This not only advance the curriculum transformation agenda to the top management list, but also ensures buy in from the broader academic community and related stakeholders. Arguments in support of formulating a monitoring structure, vision and strategy emanates from the Constitution of the Republic of South Africa where it has been contended that: "attention should be paid to improving not only access to education but also to the quality of education. This is particularly important with regard to teacher training, the curriculum, access to information for parents and learners…" (Gumede & Biyane, 2016; p69). By focusing only on access instead the guiding moral standards that govern diverse knowledge systems there are high possibilities of being trapped in coloniality web that is sustained and promoted. A detailed discussion focusing on the significance of ethics in any knowledge system appears in Mthembu (2018b, pp 26-37).

Fatoki and Chindoga (2011) suggested the significance of considering human issues in designing curriculum and skill relevance to consider human centered education in an organization and its environment. Clearly a lack of human centered curriculum result in increasing the probability of occurrence of adverse events. Such events could either be a consequence of disgruntled learners as we have seen in the recent student protest under the auspices of #*Fees Must Fall* and # Curriculum *Change* or merely an opportunity being exploited. Lack of commitment and inadequate resources distribution has also been considered as precursors of an unethical environment, thus making an organization vulnerable to a delinquent behaviour. Since most academic institutions in learning sphere are characterized by such difficulties, the importance of establishing relevant ethical environment within an organization cannot be overstated.

In proposing solutions to lack of information on moral standard that guides indigenous knowledges systems in the learning sphere arising because of inability of the narrow cultural approach to appreciate human factors, Mukuka (2010) calls for establishing normative controls. Normative controls are a by-product of a dominant security culture, which is the totality of patterns of behaviour in an organization that contribute to the protection of information of all kinds. A lack of a security culture results in problems of maintaining integrity of the status quo in the organization and indirectly threatens the protection of the possibilities of inclusion of other-previously marginalised knowledge systems. Most adverse events can be traced back to a lack of security culture and a consequence of breakdown or delays in in the implementation of the transformation agenda when it comes to structure and tuition content. As Keet (2020) highlight, besides personal factors, work situations and opportunities available allow individuals to engage in unnecessary debates. Evidence to this contention appears in Verhoef and Kruger (2020), where its shown that the prevalent work situation and the opportunity to participate in uninformed debates in various academic institutions affected the primary belief system of perpetrator, thus resulting in delaying tactics and debates being performed. The ability to leverage work situations and opportunities to engage in curriculum change suggests that monitoring of employee behaviour is an essential step in maintaining the integrity of an organization. Such monitoring does not necessarily have to be formal and rule based. In fact informal monitoring, such as interpreting curriculum behavioral changes and identifying personal and group conflicts, can help in establishing adequate checks and balances.

In the previous section the problems with establishing structures or responsibility was mentioned. Clearly adopting adequate structures will go a long way in establishing good management practices and will set the scene for effective implementation of multicultural education system. The notion of structures of responsibility goes beyond the narrowly focused concerns of specifying an appropriate organizational structure. Although important, exclusive focus on organizational structure issues tends to skew the emphasis towards formal specification. Asante (1987) introduced the concept of structures of responsibility to the agency literature. And suggest that such structures provide a means to understand the manner in which responsible agents are identified within the context of the formal and informal organizational environments. A structure of responsibility focuses also facilitate an understanding of the range of conduct open to the responsible agents, the influences they are subjected to, the manner in which they signify the occurrence of events, the communications they enter into.

The most important element of interpreting structures of responsibility is the ability to understand the underlying patterns of behaviour. The positive connotations of interpreting behavioural attributes in developing and designing secure environments has been well documented (e.g. refer to the studies conducted by Fatoki & Chindoga, 2011; Adugna, 2006; Mthembu, 2005).

Besides focusing on formalized rule structures and establishing an adequate understanding of ethical standards and related knowledge practices, it is also important to develop and implement relevant policy. Capability and relevance are a key to the design of technical control measures. In the literature there are a number of approaches available. And most of these have been tested for their validity and completeness. A challenge, however, exist when academia ignores value systems of other knowledges. Obviously in the learning environment the formal models for learning fall short of maintaining their completeness inclusivity and validity.

ORGANIZATION OF THE BOOK

The book is organized into nine chapters. A brief description of each of the chapters follows:

Chapter 1 identifies the existing challenges in the academia after two decades after scraping of apartheid policy, that was notorious of marginalisation of indigenous knowledge systems in the higher learning in South Africa. The chapter sets the scene for discussions presented by various authors. In particular, the chapter identifies the orientation of the present education system and the related problems with managing diverse knowledge systems and curriculum development. It also identifies the importance of establishing relevant curriculum, structures of responsibility and relevant knowledge recovery plans.

Chapter 2 outlines the goal for transformation that embrace different cultural backgrounds. For ethical research consideration approaches to embrace African context, there should be transformation of learning environments and institutional cultures in the education system, given the continued prevalence of upsetting and exclusionary elements that are keenly maintaining colonial exclusion measures.

Chapter 3 establishes the need for introduction of indigenous knowledge systems in the learning sphere and presents a sample structure that may be used to develop such curriculum. The author of this chapter contends that by investing in the development of indigenous - local knowledge systems, in the academia that will ensures the highest level of scholarship against all forms of pseudoscience.

Chapter 4 takes philosophical orientation and debates about the rights and wrongs in the information age. The author examines some challenges in ethical management of information technology resources. The overall aim of the chapter is to consider moral issues pertaining to computer use and misuse and articulate methods of thinking through various concerns.

Chapter 5 reviews the synthesis of principles of constructivism with the political and legal dimensions of educational pluralism and its significance in designing a curriculun that tackels socio-political injustice.

Chapter 6 reviews issues the ethical elements of education system such that cultural values of African values such as indigenous African languages could be promoted in learning sphere. The authors argue that African values systems raises confidence and hence transformation agenda, which is so important when engaging in revision of curriculum in the post-colonial era in South Africa. They further suggest that importance be paid to developing ethical policies.

Chapter 7 analyses and compares recent approaches for development of a transformed classroom. The author systematically reviews the philosophical assumptions and presents gaps and problems in each of the current approaches. A systematic position for future curriculum and practice is then established.

Chapter 8 presents analyses that suggests a multicultural approach for development of curriculum that reflect the transformed education system in the post-colonial or post-apartheid era. The author systematically discusses the philosophical assumptions and presents gaps and problems in each of the current approaches. A systematic position for future curriculum and practice is then established.

Chapter 9 concludes that the teaching of yoga and its relevant in today's societies in relation to challenges, scientific engagement and medical research. The author argues that it is possible to include yoga in the curriculum and hence it facilitates healthy life style, importance is placed on morality measures. The authors present an array of technical tools and techniques that help in achieving this purpose.

Ntokozo Mthembu
University of South Africa, South Africa

REFERENCES

Adugna, G. (2006). *Livelihoods and survival strategies among migrant children in Addis Ababa* (MA Dissertation). Norwegian University of Science and Technology (NTNU).

Asante, M. K. (1987). *The Afrocentric idea*. Temple University.

Bangura, A. K. (2012). From Diop to Asante: Conceptualizing and Contextualizing the Afrocentric Paradigm. *The Journal of Pan African Studies*, *5*(1), 103–125.

Cajetas-Saranza, R. (2015). Approaches to Integrating Multicultural Perspectives in PNU-Mindanao: Basis for Multicultural Education Program. *International Peer Reviewed Journal, 16*, 37–55. doi:10.7718/iamure.ije.v16i1.1122

Do Vale, H. F. (2016). Explaining Education Reforms and Decentralization in Brazil, South Africa, South Korea, and Spain. *Lex localis - Journal of Local Self-Government, 4*(3), 591–612. doi:10.4335/14.3.591-612(2016)

Fatoki, O., & Chindoga, L. (2011). An Investigation into the Obstacles to Youth Entrepreneurship in South Africa. *International Business Research, 4*(2), 161–169. doi:10.5539/ibr.v4n2p161

Garriott, P. O., Reiter, S., & Brownfield, J. (2016). Testing the Efficacy of Brief Multicultural Education Interventions in White College Students. *Journal of Diversity in Higher Education, 9*(2), 158–169. doi:10.1037/a0039547

Grove, J. (2017). *Black academics soon to outnumber white scholars in South Africa: Tipping point likely to be between 2020 and 2025, new demographic analysis suggests*. https://www.timeshighereducation.com/news/black-academics-soon-outnumber-white-scholars-south-africa

Gumede, V., & Biyase, M. (2016). Educational reforms and curriculum transformation in post-apartheid South Africa. *Environment and Ecology, 7*(2), 69–76. doi:10.21511/ee.07(2).2016.7

Higher Education South African. (2014). *South African Higher Education in the 20th Year of Democracy: Context, Achievements and Key Challenges HESA presentation to the Portfolio Committee on Higher Education and Training Cape Town*. Retrieved from https://pmg.org.za/committee-meeting/17037/

Iheriohanma, E. B. J. & Oguoma O. (2010) Governance, Leadership Crisis and Underdevelopment in Africa: An Explorative Discourse. *European Journal of Social Sciences, 12*(3).

Kallaway, P. (2002). *The history of education under apartheid, 1948–1994: the doors of learning and culture shall be opened*. Pearson.

Keet, A. (2020). University transformation at the crossroads. *University World News, 260*. Retrieved from https://www.universityworldnews.com/post.php?story=20200225094803518

Madiba, M. (2014). Promoting Concept Literacy through Multilingual Glossaries: A Translanguaging Approach. In L. Hibbert & C. van der Walt (Eds.), *Multilingual Universities in South Africa: Reflecting Society in Higher Education* (pp. 68–87). Multilingual Matters. doi:10.21832/9781783091669-007

Mona, M., & Hurst, E. (2017). “Translanguaging” as a socially just pedagogy. *Education as Change*, *21*(2), 126–148. doi:10.17159/1947-9417/2017/2015

Msila, V. (2007). From Apartheid Education to the Revised National Curriculum Statement: Pedagogy for Identity Formation and Nation Building in South Africa. *Nordic Journal of African Studies*, *16*(2), 146–160.

Mthembu, N. (2005). Survival strategies of the individuals and households affected by unemployment in eThekwini Municipality. In S. Benjamin (Ed.), *From the depths of poverty: Communities in post-apartheid South Africa* (pp. 1–21). University of KwaZulu-Natal.

Mthembu, N. (2018a). Ushering transformative change: Infusing decolonised cultural change into South African curriculum and andragogical practices. In D. P. Peltz & A. C. Clemons (Eds.), *Multicultural andragogy for transformative learning* (pp. 194–206). IGI Global.

Mthembu, N. (2018b). Information and Communication Technology (ICT) and Its mixed reality in the learning sphere: A South African Perspective. *International Journal of Virtual and Augmented Reality*, 2(2), 26–37. doi:10.4018/IJVAR.2018070102

Mukuka, G. S. (2010). *Reap what you have not sown: Indigenous Knowledge Systems and Intellectual Property Laws in South Africa*. Pretoria University Law.

Owusu-Ansah, F. E., & Mji, G. (2013). African indigenous knowledge and research. *African Journal of Disability*, *2*(1), 1–5. doi:10.4102/ajod.v2i1.30 PMID:28729984

Phaahla, P. (2014). Indigenous African Languages as Agents of Change in the Transformation of Higher Education Institutions in South Africa: Unisa. *Nordic Journal of African Studies*, *23*(1), 31–56.

Ramdass, K. (2015). Leadership challenges: A higher education perspective. *Proceedings of PICMET '15: Management of the Technology Age*, 1112 – 1120.

Republic of South Africa. (2019). *Protection, Promotion, Development and Management of Indigenous Knowledge Act, 2019. Government Gazette No. 42647* (Vol. 650). Pretoria: Government Printer.

United Nations. (n.d.). *Universal Declaration of Human Rights*. Retrieved from https://www.un.org/en/universal-declaration-human-rights/

Verhoef, A. H., & Kruger, H. A. (2020). What connects Shaka Zulu, decolonisation and maths models? *University World News, 260*. Retrieved from https://www.universityworldnews.com/post.php?story=20200225094803518

Zuze, T. L. (2012). The challenge of youth-to-work transitions: An international Perspective. In H. Perold, N. Cloete & J. Papier (Eds.), Shaping the future of South Africa's youth: Rethinking post-school education and skills training (pp. 42-59). Cape Town: Centre for Higher Education Transformation (CHET).

Acknowledgment

Thanks to authors who contributed their chapters in this publication of which enable the accomplishment of this project. Special thanks to IGI Team and Reviewers for their continuous unwavering support in this instance. This project was conducted within the framework of Izimfundiso zaseThebes – Africa Project at University of South Africa, Department of Sociology, School of Social Sciences and no financial support received in this instance. Ethical Clearance number: 2019_SOCDERC_002
The author declares no competing interests.

Introduction

This publication is composed of compilation of chapters that are cutting-edge scholarly resource that examines forthcoming methodologies and strategies on educational reform and the updating of curricula to accurately reflect cultural shifts. The book examines the bias and problems that prejudice creates in educational systems around the world that have been dominated by Western forms of knowledge and scientific processes. Individually and collective authors have experience and scholarship within the areas of indigenous knowledges systems, ethics, adult learning, human capital development, self-directed learning, autonomous learning, e' learning – learning through use of information and technology and andragogy. The construct of related moral standard as a transformative instrument is at the forefront of all of the chapters in this compendium which, Ngugi wa Thiong'o defined in several texts during the 1960s (Mwesigire, 2017).

Transformation is normally defined as an idea with multifaceted, open-ended and indistinct that it is basically an impracticable term. Nonetheless, the etymology of the word can be linked to Late Latin transformation (noun) rooted in the Latin verb *transformare*. In other words, the prefix trans meaning 'across', 'beyond', 'above', or 'bridging' is combined with *formare* meaning 'to form' (Du Preez, Simmonds & Verhoef, 2016:1). In the late 1500s, the notion of transformation was used to refer to 'undergoing a change in form'. In other words, to transform denotes to cause or undergo a change in forma, nature or appearance. Furthermore, transformation can also describe processes of change in physics - change of one element into another, mathematics - conversion of one figure into another of a similar value, linguistic – conversation of one syntactic form into another, biology – amendment of a cell, personality – alteration of character or appearance and politics – philosophy driven change. Yet again the Greek word for transformation, *metaschimatismos*, stresses change in form and also include the notion of remodeling, modification and restructuring (ibid.).

The Greek term for transformation, *metaschimatismos*, emphasises a change in form, but includes the idea of 'remodeling, modification, and restructuring (Du Preez et al, 2016).'. The noun 'transformation' is indeed complex and open-ended, but its core meaning is 'to undergo and/or cause a change of form'. In simply terms, the idea to 'transform' is linked to change and cannot be associated with rigidity since it depicts constant remodeling, modification and restructuring that encompass internal or external processes. The idea of transformation has assumed great importance especially when it comes to understanding progress made thus far in the demise of global colonialism and apartheid policy in South African context.

This is because the term has been used as a descriptor that is attachable to almost any phenomenon associated with social change. Conceptions of change can be, and are, associated with relatively unobtrusive interventions as much as with wholesale social engineering. They refer to both minimalist and maximalist interpretations of change, to politically "conservative", "liberal" and "radical" constructs of social intervention, to processes, to the means of change and to their purported outcomes. Transformation is called upon to evaluate post-apartheid progress in every realm of life and in areas touching on such varied subjects as affirmative action, institutional culture, the uses of language, the "colour" and gender of the judiciary, the structure of the bureaucracy in the civil service, nationhood, values and socialisation, administrative and managerial systems, institutional outputs, legislative policies, governmental practices, plans, strategies, and their consequences, and so on.

This scenario invites the need for greater clarity, certainty and definition since the elusiveness of the concept of transformation is analytically untenable. For the purposes of the book it is necessary to provide a framework for thinking about what is implied in the uses of the concept of transformation and what is not in relation to African ethical standards. The writers in this book are concerned about the limited theorisation and understanding of the concept of ethics, since it is hoped that such a theorisation will serve the purpose of informing readers of the report about how the concept itself might be understood in the wider discourses in education and training in South Africa, and even outside the immediate ambit of educational discourse.

Context, too, is important in the way in which transformation is approached by the reports. They endeavour to provide a context for the evaluation of the project of transformation, and this is related in the 2001 report to the discourse of policy and implementation. Important contextual factors affecting the development of education and its transformation refer to a wide range of national and global issues. These include the international environment in which the transformation takes place within the South African national setting and, in regard to these global forces. However, the past structure of the differentiated and unequal nature of the higher education sector in South Africa was also reflected in the differentiated or varying experiences of institutions and their manner to deal with restructuring and curriculum change.

The development and acknowledgement of relevant education curriculum that incorporates community learning and experiences including ethical guidelines for a multicultural education system through a particular disciplinary lens becomes a critical resource that will enable people to secure their livelihood in general. Presently, the education landscape especially the learning and teaching theories have been centered on the narrow cultural settings. However, the post-colonial era has tended to facilitate the notion of agora [the public space in which science and the public converge and the public interact with science] in the transformed education system.

This book provides a valuable window on information assurance and covers the necessary components from challenges and possibilities for indigenous knowledge systems in the learning sphere. Understanding of social order and related ethical codes are a critical aspect to avoid misrepresentation and misinterpretation of knowledge, thereby perpetuating diverse knowledge and advanced scholarship. They must be configured with a set of filtering rules and must be constantly altered to address new challenges and possibilities. Currently, vast literature tends to ignore ethics when dealing knowledge creation and general scholarship. Because of its uniqueness, value systems, such as lingua franca, moral standards are becoming a promising means of authentication. Ngugi wa Thiong'o has elaborated in the historic writers' Conference held in Makerere University, Uganda in 1962 when he stated that the authenticity and a return in writing in indigenous African languages, avoiding imperialism and subjugation of the colonizers (Oniwe, 2017). In order to ensure the current literature to be relevant it's significant that it should be created and documented in languages that are African languages to qualified to be regarded as an African literature. Nonetheless, its objective is not to undermine other knowledge systems, the dominant, but to provide the diverse and methodology by which multicultural knowledges could be managed.

The challenges regarding transformation of education are both trying and interesting. People are working on them with enthusiasm, tenacity, and dedication to develop new methods of analysis and provide new solutions to keep up with the ever-changing learning scenery. In this new age of global interconnectivity and interdependence, it is necessary to provide scholars, both professionals and learners, with state-of-the art knowledge on the frontiers of knowledge systems. This book is a good step in that direction.

Ntokozo Mthembu
University of South Africa, South Africa

REFERENCES

Du Preez, P., Simmonds, S., & Verhoef, A. H. (2016). Rethinking and researching transformation in higher education: A meta-study of South African trends. *Transformation in Higher Education*, *1*(1), a2. doi:10.4102/the.v1i1.2

Mwesigire, B. (2017). *What happened to Ngugi wa Thiong'o's Makerere University?* Retrieved from March, 10, 2020, from https://thisisafrica.me/arts-and-culture/happened-ngugi-wa-thiongos-makerere-university/

Oniwe, B. A. (2017). *Narrating the (Im)Migrant Experience: 21st Century African Fiction in the Age of Globalization* (Doctoral dissertation). Retrieved from March, 13, 2020, from https://scholarcommons.sc.edu/etd/4524

Section 1

Novelty of Indigenous Knowledge

Chapter 1
Fighting for Relevance:
The Revitalization of African Knowledge in the Learning Sphere in South Africa

Nombulelo Tholithemba Shange
University of the Free State, South Africa

ABSTRACT

South Africa's recent higher education protests around fees and decolonizing institutions have shone a spotlight on important issues and have inspired global discussion. We witnessed similar resistance during apartheid, where African languages and ideas were limited. The educational space was the most affected by clashes between languages and ideas; we saw this in the prioritizing of English and Afrikaans over indigenous African languages and the prioritizing of Western medicine, literature, arts, culture, and science over African ones. This chapter will show how formal education and knowledge production in South Africa has been used as a tool to repress Black people, while discrediting their knowledge systems. This discussion will draw from impepho, which is rejected by Christians because its main use is for communicating with ancestors. The herb has many other medicinal uses, but it is still rejected. African practices are used and revitalized by AIC like the Shembe Church and revolutionary movements like FMF.

INTRODUCTION

African knowledge systems, science, philosophy and culture have been sidelined from mainstream educational institutions. African inventions and uses have largely been written out of history and left out of school syllabuses. This has been done

DOI: 10.4018/978-1-7998-1249-4.ch001

through the use and abuse of power; Western settlers decided what information was valuable and should be taught. Historically the driving factor has been the Western mission church, which pushed the agenda of the inferiority of Africans and alienated people from their culture for colonial and capitalistic reasons. The "othering" and sidelining of all that is African was perpetuated through vulgar and ethnocentric racism, which Frantz Fanon discusses as having taken place in stages. Bantu Biko believes that this racism has gone as far as instilling self-hatred in black Africans, an issue that has been a huge hurdle in South Africa's development and in the development of black identity as a whole.

Racism as a controlling tool worked its way into schools and was used to destabilize any system that challenged the Western Christian Church, capitalism or white privilege. Historically this can be seen through the repression of African Independent Churches like the Shembe Church, where missionaries and the state used educational systems and the media to push false propaganda, which put members at odds with schools and earned members the unfair reputation of being backward and uneducated.

The 1976 and more recent #FeesMustFall student protests further showed the oppression directed at black students, African knowledge systems and culture. Students have called for decolonized education that puts more relatable African ideas first, in the hope of restoring cultural identity and dealing with developmental challenges. Indigenous knowledge, like the uses of important herbs like Helichrysum or *impepho*, remain largely unutilized, despite its functions, accessibility and medical benefits. This chapter will discuss these themes while highlighting the benefits and challenges to decolonizing educational institutions and using indigenous knowledge as a tool to do so.

METHODOLOGY

The discussion on decolonization and the relevance of indigenous knowledge is not one that should only be explored on a curriculum or content level. We should be mindful of it even in teaching methodologies and how we conduct research and produce knowledge. This research will use qualitative methods. Qualitative research is unstructured, flexible and relaxed and aims to describe and explain a situation, phenomenon, problem or event (Mason, 1994: 101). This method, while Western, does in some ways lead us closer to decoloniality because it gives the participants room to shape the narrative and the direction the research should take and allows for the inclusion of rich history and background the researcher might have overlooked. It is important to consider the historical background of the research topic, people's opinions of and responses to issues and detailed descriptions of research issues

(Kumar, 2005: 12-13). The central research technique was in-depth interviews. This paper drew from secondary sources such as books, articles and online sources It also relied on ethnographic research done on the Shembe Church. The research was done as part of Master's degree requirements between 2011 and 2013. The thesis is titled: *Shembe Religion's Integration of African Traditional Religion and Christianity: A Sociological Case Study*. The bulk of the primary research was conducted in eBuhleni, one of the biggest Shembe factions. eBuhleni is an Inanda Shembe Village situated in Kwa-Zulu Natal (KZN), where gatherings take place in July and September each year. The researcher visited this division of the Shembe Church during their July gathering in 2011, where I documented details about the Shembe history, ideas, practices and the followers' perceptions of how they are treated individually and institutionally as a religious group. More information was gathered through visits to ekuPhakameni, the oldest faction of the Shembe Church. I also visited Shembe Temples in and around Kwa Mashu and Inanda in KZN. Interviews and focus group discussions were conducted with Shembe followers and strategic informants in their homes, places of work and temples. More interviews were conducted to explore some of the themes of this paper, which were not fully covered in the Master's research. Two traditional healers were interviewed; namely Buhlebendalo Mda, who is also a musician and runs a mentorship programme, and secondly, Tabiso Gobela, a Witwatersrand University graduate, who also partook in Fees Must Fall protest action. He now works as a sangoma and a TV producer. The idea was to speak to healers, and those who possess indigenous knowledge. In-depth questions were asked around the usefulness of the knowledge they pose, which they believe should be mainstreamed and taught in institutions of learning. Finally, I interviewed student leaders and students who were involved in Fees Must Fall (FMF) action and the continuing decolonializing projects taking shape in many institutions all over South Africa. The participants were Lehakoe Masedi, teaching and learning facilitator and Master's student at the Law faculty at the University of the Free State (UFS). I also spoke to Mosa Moerane, an LLB student who also works as a student assistant at the UFS Centre for Universal Access and Disability Support. Finally, I spoke to Busisiwe Ntsele, a PhD candidate and former lecturer at the UFS Sociology department. Busisiwe did her undergraduate studies and her Master's in the law department at the University of Johannesburg. These interviews focused on reimagined educational institutions and what decolonized structures might look like, as well as the challenges in making such structures a reality.

HISTORY OF REPRESSION: THE RISE OF WESTERN RELIGION, EDUCATION AND CAPITALISM

The 1976 Soweto Uprising was one of the many turning points in apartheid resistance. It plunged the issues facing black youth into center stage, shining a light on the brutal apartheid government's reaction to resistance and questioning. This uprising challenged the damaging narrative of dysfunctional black youth that existed then and which continues to exist. The youth of 1976 took to the streets to challenge the tightening of the 1953 Bantu Education Act (Ndlovu, 2006: 318). The 1953 Act reduced funding for black schools and universities, and was designed to steer black students towards unskilled labour (Union of South Africa, 1953: 263). The 1970s amendments saw Afrikaans being introduced as the main language in schools. Students feared this move would create even more challenges in an already problematic education system. Many did not understand Afrikaans and questioned the move. Afrikaans was also viewed as the oppressors' language, making the change highly contentious (Hyslops, 1989: 10).

The Bantu Education Act was not the first time that the state used education against black people as a controlling power tool. The manner in which mainstream, Western education was delivered to black South Africans is problematic in its own right. One of the ways that western education was imposed on Africans was through religion. Africans were educated on Western Christianity, while manipulated into rejecting their own cultural and religious practices. From the moment missionaries set foot in Africa, South America and Asia, their aim was to indoctrinate the indigenous people and to manipulate them into adopting Western ideas and Christianity, which would further Western domination and strengthen Western economies. The missionaries often turned to morality and religious dogma when trying to control colonies. Many of the ideas the missionaries spread were based less on religious ideas and more on Western cultural ideas and their interpretation of the religion (Dougall, 1963: 60). One of these was Calvinism, which encouraged determinism and pushed the idea that heaven and eternal peace, are reserved for the select few, namely the rich land and factory owners. The poor were viewed as ungodly, and could only come closer to godliness through constant labour. This view led to the rise of capitalism in the West, and defined idleness as a sin (Tarsons, 1958: 101 & 103). As time went on these ideas spread by missionaries also allowed white business owners to create cheap, non-competitive workforces made up of oppressed black people (Christie and Collins, 1984: 163). These ideas continued long after the arrival of missionaries and were promoted in schools, churches and various other social structures. Many of these ideas were even used to prolong oppressive regimes such as apartheid in South Africa. The Western missionaries believed that their interpretation of Christianity was the "correct" and "universal" interpretation (Ani, 1994: 517). Indigenized versions of Christianity and African Traditional Religion were rejected; those who partook in these practices were often accused of being devil worshipers. White missionaries wanted Africans to reject their traditional beliefs; they often portrayed them as being fundamentally at odds with Christianity (Pauw, 1963: 34). African converts were therefore encouraged or coerced to reject their ancestors, cultural ideas and rituals.

The hidden agenda behind Mission Christianity as already mentioned was to allow the West's control over colonies, land and labor. Indigenous people were convinced that being idle was a sin and the only way one could prove one was not guilty of this sin was by working in businesses owned by the colonizers (Christie and Collins, 1984: 163). A "good Christian" would work in a mine or on farms, in shops, factories, white-owned homes and businesses, because whiteness was seen as being next to godliness, and Jesus was/is portrayed as white (Ani, 1994: 219). Eurocentric ideas and practices were also portrayed as "objective" or "scientific" universal discourse which could not be disputed or questioned. This in part was to sideline African ideas and create doubt around their worthiness, but also in order to further Western capitalist and political agendas centered on domination and conquering (Ani, 1994: 515 - 516). Many people gave up useful African practices and knowledge systems in exchange for earning a measly wage which was often insufficient to maintain whole families. These ideas were often spread in schools and churches. People were made to believe that they struggled because blackness was inferior, less than human or a sin (Ani, 1994: 282). They were also convinced that they struggled because they were lazy, and God only rewarded those who worked hard. Those who did work hard and still struggled nonetheless were made to believe that this was God's will or that they were not working hard enough (Douglas, 1979: 51). The Western mission often used the idea of religious determinism in order to maintain their political and economic hold and control. Determinism assumes that there are external factors acting upon an individual; the idea is that the external factors are active while the individual is passive and powerless (Douglas, 1979: 51; O'Connor, 2010). Missionaries convinced Africans that this external force was God and that they were there to carry out God's will. God's will could not be questioned because the missionaries convinced Africans that God was all-knowing and all-powerful (Zagzebski, 2004: 1). This idea still exists even today, as people going through hardship seek comfort in the idea that it is "God's will", and people will often encourage people going through difficulty by saying; "Everything happens for a reason", instead of working through solutions and questioning challenging situations. Africans then start to believe and accept that theirs is a life characterized by struggle, or that the only way they can come closer to happiness and freedom is by rejecting all that they are and subscribing to the norms set out by white missionaries and authorities, who have positioned themselves as the authorities on God's will.

DISCREDITING AFRICAN IDEOLOGY AND HISTORY: A SHEMBE CASE STUDY

Emerging African churches questioned the power missionaries gave themselves at the expense of African ideas. One such group was the Shembe Church. The Shembe Church was/is one group that was/is seen as 'amaqaba', meaning primitive and uneducated, because of the steps the church has taken to revitalize African culture, which has historically been viewed as backwards. Again, these are ideas that were spread in churches and schools by white missionaries heading up those structures. Repression within the educational system started long before the Soweto Uprising or Bantu Education, and the Shembe Church also experienced oppression at the hands of Western education and knowledge production structures like the Christian Church and media.

Many believe the Shembe Church started in 1910 (Vilakazi, Mthethwa and Mpanza, 1986: 43). The rise of African Independent/Indigenous Churches (AIC) was fueled by the desire to interpret the Bible in a way Africans could understand and apply in their daily lives. The aim of this and other African churches was to revitalise African culture, religion, knowledge systems and family structures which had been dismantled and destroyed by colonialism and industrialisation. The birth of the Industrial Revolution in SA was characterized by chaos, disorder and confusion. These changes were evident in the social, political and individual/personal spheres, and were controversial and contested. The changes were expansionist, exploitative and consumptive; they were legitimated through the ideology of progress, development and growth. The new changes brought about by technology and industrialization encouraged inequality and created huge gaps between the rich and poor. These changes also encouraged the exploitation of people and the environment for profit and growth (Jenks, 1993: 7). Many of the challenges that plagued South Africa then still exist now, over a century later. For this reason, the Fees Must Fall activists' call for decolonization and the embracing of African ideas is so important and urgent, because if left unchallenged, the status quo will remain despite the social and environmental destruction that has been caused over the centuries. Lehakoe questions the manner in which the capitalist agenda is given importance over and above the environment, despite the destruction the market has created.

*That's the excuse often used, questioning what is the use of this *indigenous knowledge* if you can't apply it in the real world? I think the way Africans think is sustainable... So sure we can do all this work *calling for decolonised education* in the university, but it also needs to be done in the world because capitalism has ruined literally everything; the world after 2030 is not going to be able to reverse climate change and that is because of capitalism, societies are becoming more* *and more unequal because of capitalism. A lot more of this discussion and steps in decolonising needs to go out into the world... Because there will come a time when even individuals with capital realise that this is not working, maybe we have to change it. Then it will be individuals who have that knowledge *indigenous knowledge* that's the kind of role we are going to have to play to show how things can be done differently. (Lehakoe, interview with author, 2019)*

The Shembe church in many ways emerged as a reaction to the spread of capitalism in SA, critiquing the way black people were exploited and used as cheap labour to service the needs of the white and questioning the associated damage to the environment (Griffin, 1995: 11). Shembe Church founder Isaiah Shembe's ideas emerged in the early 1900s, after a century of social transformation, domination and fragmentation of the African people through internal cultural warfare, colonization, Western missionaries, increased industrialization and environmental factors of drought and disease. The Shembe community provided religious, economic, cultural and moral space for early followers, most of whom were marginalized: young girls, orphans and women (Johns, 2004: 403). Isaiah Shembe and his followers created a discourse of 'cultural truth' which was set to challenge the ideologies of the state and Christian missionaries (Muller & Buchanan, 1994: 45). Song, dance and dream narratives were seen as cultural treasures and were used to convey the message of cultural truth which centered around self-acceptance and love of one's physical appearance, culture, identity and so on. The performance and cultural messages embedded within it are collectively re-enacted by the social body. These cultural performances serve a number of purposes; they communicate both daily and historical experiences of Isaiah Shembe and his followers, and this is done through written and spoken language (Muller & Buchanan, 1994: 47). They were also an important way of creating and sharing new knowledge and experiences and revitalising old ones.

Shembe's aim moved beyond restoring African culture; his movement also hoped to create unity and multiculturalism by incorporating different African, Indian and Arabic cultural practices and Western Christianity into one religious ideology. It is also this multiculturalism that needs to be explored and implemented in coming up with decolonial strategies. The Shembe Church's move towards multiculturalism has seen its fair share of rejection. In some ways, the move towards multiculturalism created even more division between the Shembe Church and anti-Christian/ anti-Western movements. The Shembe Church and many other movements were rejected by African Traditional Religion(s) (ATR) and Christianity. African religious groups rejected the Shembe Church and other churches because many were founded during a time when black people were being oppressed by Christian missionaries. Adopting Christianity therefore seemed like a betrayal. African religious groups felt that adopting Christianity weakened the anti-Western oppression resistance.

African Traditional Religionists like Mndende (2009: 1) feel that many African religions only incorporated Christianity and other Ibramic faiths to "justify" their own religion to the sceptics.

Christian missionaries and the state still felt threatened by the Shembe Church despite the fact that the church did not completely reject Western teachings and Christian ideas. An elaborate propaganda campaign was created; the story of Isaiah Shembe flying off a mountain to his death was strategically concocted. This story was created to prove that Isaiah Shembe was an impostor and a fake; this would therefore discredit the whole religious movement. According to the story, Isaiah Shembe told his followers to go to Mount iNhlangakazi (this is the sacred mountain where the annual January pilgrimage takes place) where they would see him fly to heaven. He allegedly attached feathers to his back and when he jumped off the mountain, he hit the rocks and died (Meersman, 2012). Shembe's actual death was caused by salt fever; his body was reabsorbing its own perspiration. Isaiah Shembe's son, Johannes Shembe told Fernandez (1973: 44) that his father literally worked himself to death. Isaiah Shembe often walked very long distances to baptize new members of the church. Isaiah Shembe often shared the details of how he was going to die. Johannes Shembe mentioned that he never believed his father when he spoke of how he would die. Johannes Shembe mentions that his father was so strong and healthy; he never believed that his father could die from an illness like salt fever (Fernandez, 1973: 44). The story of Shembe flying off a mountain was printed in newspapers, told in churches and taught in schools as part of South African history. The story created a lot of conflict and uncertainty within the Shembe Church as many followers were unsure whether it was true or false. The everyday insults and discrimination led many members to leave the church or drop out of school. Education was then seen as being in conflict with the Shembe religion and many of the youth felt pressure to choose between church and an education. As a result, the church became known as the church of the uneducated (Mpanza, 1999). This is because they were not receiving formal education, while their cultural and indigenous knowledge was seen as backwards or inadequate. This example is just one illustration of how colonial forces distorted or lied about African history and discredited knowledge systems. They did this in ways that were racist and that brought Africanness into question. This kind of overt oppression and racism, while no longer as prominent, now exists in more covert ways. Mosa (interview with author, 2019) raises this issue in relation to African languages being marginalised in academia and the non-verbal way in which universities communicate that blackness is not welcomed;

*The university is very interesting in the value it gives you based on different things, so you are valuable it you can speak English and you are most likely to succeed in a university like this one *UFS*, you are even better if you can speak Afrikaans.*

*It's very possible that you could get your degree just off those two things... We are a majority black country and in this university it is not hard for you to go into a department and only find one person who is black in the entire department and that person might be the admin lady at best because even the admin people on this campus are often white women... The broad categorising of women means that it is possible to say we have succeeded in our transformation goals because we have the people listed in the designated groups of the employment equity act and it could remain untransformed regardless *because mostly white women are being hired*. And these are the people you interact with in the department, so even their reaction to how they speak to white Afrikaner students versus how they speak to black students is vastly different. So circling back to your question on decolonisation, it can't only be that decolonising in education is only limited to the curriculum, to the teaching, to the learning. It's a literal dismantling, breaking apart, we literally have to deconstruct the very ways in which we exist.*

FRANTZ FANON AND BANTU BIKO

Biko's theoretical ideas are the most fitting for this discussion as they help make sense of discussions on decoloniality and students often draw from black consciousness to make sense of repressive university environments and in their call for change. Biko was also influenced by ATR in his writing and activism. His ideals on black consciousness go beyond political activism; black consciousness is about the acceptance of blackness, and just like the founding principles of the Shembe Church, black consciousness encourages black people to embrace their identity, skin, hair, culture, and tradition and religion are also part of this (Sogiba, 1996: v). When looking at Biko, Fanon's ideas are not too far away, as they both conceptualise blackness and colonial analysis in similar ways. Fanon argues there is a mutual link between culture and racism, both for the people who practice racism and for those who become victims of it. The connection between racism and culture can even be explained according to the different phases of colonialism. Fanon makes a distinction between two types of colonial racism. The first is vulgar racism; here the inferiority of the indigenous people is proven by citing physiological science, with claims like those that Africans have smaller brains than Europeans (McCulloch, 1983: 119). The second is ethnocentrism; which is more sophisticated than vulgar racism. In ethnocentrism, the aim shifts from objectifying the physiology of the individual to objectifying the cultural style of a group. Vulgar racism became embarrassing after the brutal way in which it was used in Nazi Germany; it also became increasingly hard to defend in the colonising countries owing to the growing anti-colonial consciousness (McCulloch, 1983: 120). Ethnocentrism is more complex, subtle and often harder to identify. It relies more on cultural differences rather than biological differences. An example of ethnocentrism is how white Christians sidelined and rejected the cultural use of *impepho* and filled the gaps with Western culture and practices. Mosa's (interview with author, 2019) statement shows how ethnocentrism can also be used as a tool to delegitimize knowledge systems, and celebrate Western ones at the expense of African ideas;

In 2015 I attended an environmental summit and there was a white woman who was presenting, she was an academic and she had received funding from Europe to teach black people in the townships how to build mud houses. Because now they have come to realise the sophistication of the mud house, even the ones we typically build with bottles, they are now calling them eco buildings, saying they are cool in summer and warm in winter. In Sotho we call them "mokgoro", and they are our houses and it's how we live. But now suddenly because a white woman said so, they are now worth millions of Euros as an idea, as a concept. And the reason why we got into these stupid ways of concrete is because they came from Europe and said our mud houses were uncivilised.

Ethnocentrism is another discourse that the powerful use to justify domination. Stereotypes help in creating and perpetuating ethnocentrism (Blaut, 1992: 289). One can therefore conclude that ethnocentrism is broad and it manifests itself in many different ways and different fields in society. This makes it harder to identify and therefore more difficult to alter. Ethnocentrism can be an individual or group's refusal to acknowledge another individual or group's ideas and practices or structural oppression that leads to one or more groups being denied resources and opportunities based on their race, culture, ethnicity, religion and so on.

Fanon (1964: 44) believes that racism is just one way in which colonial oppression discredited people's entire way of life; ranging from their language, religion, culture, dress, food, indigenous knowledge to all aspects of society. The racist colonialists had no interest in the indigenous cultures; their main concern was to establish complete control over a given territory (McCulloch, 1983: 122). Racism was one tool that helped the colonialists achieve this goal. Fanon and Bantu Biko have very similar ideas on the subject of colonial discourse and the impact imperialism had on the indigenous people. According to Biko (2004: 54) "The system has allowed so dangerous an anti-black attitude to build up amongst whites that it is taken as almost a sin to be black". This anti-black attitude exists beyond Western society; it has been so ingrained into people's mindsets that even indigenous people (black people in the case of Biko's writings on South Africa) start to believe it. According to Biko (2004:111), the "black man" internally develops a certain state of alienation and hate. Indigenous people reject themselves and everything they stand for, from the biological level like the color of their skin and texture of their hair, to their social make up including culture, religion and language. The colonized reject their own identity because they attach the meaning of whiteness/Western values to all that is good (Biko, 2004: 111). The hatred that people often feel towards themselves and all which resembles their identity comes from the way they live and the ways in which they were socialized as children. This situation applies today as working class black people in post-apartheid South Africa are still deprived of resources, quality education, opportunities and recognition.

During apartheid, Black South Africans found themselves in a state of alienation; they found themselves feeling inferior and less than human. The inferiority which many black people felt stemmed from observing their surroundings and making comparisons between predominantly black areas and predominantly white/middle class areas. Most black people looked at their surroundings and found that their schools were not the same as white schools. The education many black people received was inferior to that of white people. The black individual saw that homes were different, streets and neighbourhoods were different, with white neighbourhoods being better and superior (Biko, 1998: 362). All this observation led the black individual to one conclusion; that something is incomplete or missing within black humanity. "The black individual then began to attribute completeness to whiteness. This idea was then carried through to adulthood and was often passed down to the next generation" (Biko, 2004: 111).

Colonial oppression is sophisticated in that it changes in order to maintain its control, which today comes in the form of less direct neo-colonialism. Kwame Nkrumah (1965: 69) believed that neo-colonialism is a kind of "recolonization". Neo-colonialism represents the last and most dangerous stage of imperialism. Like ethnocentrism, it is complex, sophisticated and harder to identify (Nkrumah, 1965: 69; Hook, 2004: 87). Colonial oppression is always changing its face, from colonialism to neo-colonialism, vulgar racism to ethnocentrism. The more colonial oppression changes, the harder it is to identify, making it harder to eliminate. Colonialists even go as far as creating differences between people; this way they are able to maintain control while people quarrel amongst themselves (Posnock, 1997: 327). Biko (2004: 60) states that people were divided into two camps- the converted "amagqobheka" and the pagans, "amaqaba", which also means uneducated or backwards. Members of the Shembe Church were categorised as belonging to the latter camp.

SOLUTIONS: RECLAIMING OLD AFRICAN IDEALS

One of the ways that Western society gained control of Africans and their consciousness is through the historical and continual discrediting of important African rituals, culture and even science. "Amagqobheka" were forbidden to use *impepho* and other important herbs with many functions and uses (Pauw, 1963: 34). Buhlebendalo describes the manner in which black Christians still reject many aspects of African culture as a kind of "recolonization".

We recolonise ourselves, here's mpepho, all our grandmothers used to burn it, but when we need calming and cleansing in our lives we'll look at westernised things for solutions, you'll go to the pharmacy to buy expensive sage, leaving the mpepho you grew up with behind (Buhlebendalo, interview with author, 2019).

Impepho is an important holistic, ritualistic, cultural and medicinal herb (Becken, 1968: 141). The scientific name for *impepho* is Helichrysum, it has many uses in the African context, including burning it to awaken the ancestors when the living wish to communicate or receive messages from them. *Impepho* is also used by individuals, sangomas or traditional healers, allowing them to communicate between different realms (Zondi & Ntshangase, 2013: 230). *Impepho* can be used in various ways, from burning it, to drinking it as a tea, taken raw, in bath water, using it as an ingredient in traditional beer and many other methods (Lourens, Viljoen & van Heerden, 2008: 631). The medicinal uses are just as diverse, healing everything from colds and flu, fever, insomnia, weak libidos, allergies, diabetes, epilepsy, smallpox and hysteria, to its use for wound dressing and even pain relief. The burning of the plant is also said to help clear chest infections (Lourens, Viljoen & van Heerden, 2008: 630). The versatility of this and many other African herbs is impressive. Even though a herb like *impepho* and many others can cure a whole host of illnesses in ways that few Western medicines can, it is still sidelined or used as an "alternative" form of healing. Much of the #FeesMustFall student movement calls for a decolonized South Africa and decolonized educational structures, which put African science, knowledge and philosophies first. The call for decolonized education challenges our current reality. Black South Africans have historically been isolated from their land and its functions, and while African systems are being revitalized in post-apartheid South Africa, much of the old knowledge remains lost, sidelined, or is not commonly known, and is often contested and questioned for its validity and contradictions. Busisiwe (interview with author, 2019) however, sees this as an opportunity and believes that this is how new knowledge is created, through discussion, questioning and testing it.

Diversity creates creativity so contradictory African knowledge is not a bad thing. We don't have to have a singular African system... doing that will take us back to the binary systems that the West created. For us it's about saying we want to take African methodology from the periphery and start centering them around our way of life, because we do have a way of life that has worked for ages. So tapping into those things in the 21st century, for me that would be decolonised education. African ways were shifted into the periphery not because they didn't work, but people came and said that they were barbaric. And you find that even the ways that were said to be barbaric, years down the line those very same people who discredited them are using those same methodologies and now we are paying so much to get them back to Africa. We also need to start with changing our mind-sets to stop believing that Africans are inferior. When we start look at ourselves as human beings and realising that we are worth more than what the colonisers have programed us to think we are, that's when we will improve our knowledge systems in terms of knowledge creation. Because knowledge is not just everything that happens in the classroom, especially in African countries, the best kind of knowledge is the knowledge you get from outside the corners of the universities.

Busisiwe also raises important points about the way African knowledge and practice is culturally appropriated, whitewashed and sold back to Africans as though it were a new product or idea. The question should no longer be; "Is indigenous knowledge important to teach?", instead the question must be posed, "When should we start teaching it, how should it be taught, what are the obstacles and how do we address them?" Lehakoe (interview with author, 2019) adds;

*What we should be asking ourselves is whether or not there is still space for other forms of knowledge, because what I imagine when I think of decolonised institutions, is to have indigenous knowledge (IK) at the center, even if it is in the methods of teaching, if it's not in the curriculum and then having everything else around that. I think trying to squeeze it *IK* into the curriculum and formal education in the way that we have it right now is not feasible...we must change everything. The *indigenous* knowledge isn't the problem, because like we said with the mud houses, our old knowledge systems are presented back to us by the Western world. The problem is through who is presenting the knowledge, when it is presented by us.*

Here Lehakoe also makes brief mention of the way in which African ideas are sold back to black Africans by outsiders and she makes reference to mud houses, a project which is being developed in Bloemfontein, where the erasure of the African origin is clear because the homes are referred to as eco houses. The project has generated funding, but it goes to the University of the Free State because rich European donors do not trust poor black people with money. The programme also problematically draws references and inspiration from Orania, the racist whites-only community. The project has also arranged visits to Orania to see further examples, despite the fact that Lesotho is in close proximity to Bloemfontein and there are many examples of mud houses there and all over South Africa.

Buhlebendalo raises similar critiques in her earlier comment when she questions why people sometimes prefer commercialised sage, over traditional *impepho*, and in doing so, participate in the process of their own cultural erasure. While the reality remains, African solutions are the only things that can save Africa from its problems. A herb like *impepho* is cheap, holistic, easily accessible and has many medical functions. With SA's health challenges and expensive healthcare systems, it would be a huge advantage to have the knowledge of the medical uses of this herb more commonly known and encouraged by medical practitioners, government, researchers and society. Additionally, it would be useful to create interdisciplinary discourse around something like *impepho* between the traditionalists, sciences and social sciences. The traditionalists can provide valuable information on its cosmological and health uses, which the health scientists can research further, and the representatives of the social sciences can analyse the way this herb creates better access to healthcare, unity and social cohesion from its use in rituals.

SOLUTIONS: NEW STRATEGIES TO CONSIDER

Much of what needs to be done in order to decolonize the curriculum is already being attempted in many ways, so there is no need to treat this process like an unachievable one. It is difficult to prescribe how this should be done; in fact, we should avoid prescribing or forcing certain agendas, especially because there is no one way to decolonize and we often differ on solutions. Take Karl Marx for example: his discussion on the communistic revolution and the conflict between the proletarians and capitalists is one that has inspired social justice activism in South Africa for many years. But even Marx's ideas are being challenged in decolonial discourse. Marx draws much of his thought from Georg Wilhelm Friedrich Hegel who was responsible for the racialism of Germany, and some of whose ideas were also later on adopted by the Nazis (Bonetto, 2006: 35). He is further critiqued for his view that the world begins in Asia and ends in Europe, completely ignoring Africa (Bernasconi, 2001: 171). Some believe both Marx and Hegel's views should be removed from decolonized discourse, while others think it should only be Hegel who should be excluded. Many others believe we should not erase entire ideas and risk losing important historical context; instead, we should keep the positive and disregard the repressive, while still others believe even retaining the negative aspects is good, so that we never forget the violence of the West, which even made its way into intellectual discourse. These disagreements in many ways mimic the lack of cohesion between different AIC and how many questioned the Shembe Church's multicultural views that adopted the positive parts of Christianity. It is unclear how we should deal with these disagreements, which are not new to the decolonial discourse. There are many challenges linked to decolonization and they are not only linked to the recent calls for decolonization by FMF activists. In the 1960s and 1970s theorists like Kwame Nkrumah and Amilcar Cabral expressed similar concerns over decolonization, where Cabral (1969: 75) warned liberation leaders to be mindful of "ideological deficiency" and pay attention to the kinds of societies and futures they hoped to create. This is a struggle that even FMF students battled to keep in mind, and which was arguably the undoing of the movement. One of the successes of FMF was that it forced universities to take on the decolonial agenda by changing systems and curricula, but even institutions are failing at finding some form of consensus on what the goal is, what should fall and what should remain, and the agenda is increasingly being viewed as a managerial process which no longer speaks to the calls made by students in 2015 and 2016. Nkrumah (1965: 240) also warned against this, where he feared that neo-colonialism would be confused for decolonisation and taken up by actors whose true intentions are to main the status quo as much as possible. The change ends up being temporary or reforms rather than revolutionarizes, and keeps Western thought, culture, economic needs, and so forth at the core, while African systems remain in the periphery (Ndlovu-Gatsheni, 2013: 71). African systems, knowledge and practices are once again disregarded or sidelined through this process.

It is important to ensure that decolonization strategies are bottom-up participatory processes, rooted in frequent dialogue, consultation and shared experiences and strategies. These are just some of the strategies I have adopted in my work at the University of the Free State's Sociology Department. The department is said to pilot a new Social Movements course in 2021 and these are some of the strategies adopted; many of which can be readapted and applied to different contexts:

- Drawing More from African Theorists and Case Studies: The idea is to have African theories as the core of the curriculum and the course objectives see to this in a significant way. This is done by mapping social movements in Africa, their history of challenging colonialism and by looking at the following theorists: Kwame Nkrumah, Bantu Biko and Frantz Fanon. The discussion around gender discourse and movements draws from Amina Mama and Obioma Nnaemeka. Some of the case studies include local examples like service delivery protests in Bloemfontein, the work being done by Abahlali Basemjondolo. Arab Springs, Occupy Wall Street and Zapatista protests also provide a global comparative juxtaposition.
- Knowledge Exchange Week: The proposed knowledge exchange week would be a way of reimagining learning. The idea would be to invite

different individuals working in this space; from activists/ members of social movements, NGOs, academics, students and maybe even media and/or social media representatives reporting on or somehow involved with social movements. This week should help students understand different perspectives that are not just academic in nature, give a glimpse as to how what they have learnt can lead them on different career paths, show real-life experiences and deepen their understanding of the discourse. This will be done through important life histories they would not otherwise get to experience in the lecture room. The idea would be to have each of these "stakeholders" address the students during lectures. We could take advantage of them being around for a few days by also arranging seminars or panel discussions on campus, and this could even be a collaborative effort with other departments or neighbouring institutions of learning. This week can also be one of the major ways in which we provide concrete opportunities for evidence-based learning and engaged scholarship. The different views that are brought in by those working outside of the university space will bring in African, grassroots and alternative perspectives not often found in university spaces, but which are needed in these spaces in order to start the process of decolonizing them.

- No Tests and Exams: The idea is to do away with "traditional assessments" like exams and tests. Doing away with the exams and tests is another way in which this course tries to decolonize higher learning. Exams do not offer any tools or strategies students need to build on their career or help them navigate through life. Instead, they are a stressful process for both students and lecturers and they have led to a high prevalence in student suicides, depression and burnout. Innovative assessments should teach students new skills that are important for the different careers they might find themselves in, including but not limited to things like conducting interviews, engaging with different stakeholders, debate and discussion, presentation skills, journaling, and even a fun social media assessment where students can learn how to create awareness through memes, short videos, podcasts and many other social media tools. Many of these are group assessments which place an emphasis on shared learning and students supporting each other. This kind of assessment will indirectly lead to the building of other skills, such as confidence, empathy, interpersonal skills and the ability to work with others, a stronger ability to observe and make sense of one's surroundings, and how to defend one's points of view and build strong critiques around ideas they disagree with. These are all "soft skills" that are important in the workplace and in general social engagements.
- Frequently Workshopping Courses: Students often find the content taught dull, boring or irrelevant. Students often do not have a say in curriculum

> design except through problematic standardized student evaluation forms that do not capture the nuances of each course and are usually not taken seriously by either students or staff. The new Social Movements course hopes to remedy this by workshopping the course with students and other departments and possibly other universities. This will be done before the course is piloted. It is also important to perform this process repeatedly, maybe every three to five years, to ensure that the content is still relevant and engaging.

On a grander scale, these and other decolonial strategies need to have greater alignment with institutional practices. Decolonized institutions should not be institutions that exclude or deregister students for financial reasons, they should not respond with force and violence when they are challenged by students and staff through protests or other means. They should also care about students' socio-economic challenges like struggling to get to campus because they live far away or do not have transport money, struggling to pay rent or living in inadequate accommodation, students who are food insecure, and so on, because even the best and most imaginative, relevant decoloniality strategies will fail when they are offered to hungry students who cannot think beyond where their next meal will come from, or how they will get to campus, for example.

FUTURE RESEARCH DIRECTIONS

There is a growing trend in South Africa and Africa at large to reclaim lost African ideals. More and more scholars are reading and writing about them, social movements like FMF are inspired by them and more people are openly practicing cultural rituals, consulting sangomas, honoring ancestors and so on. Many Africans are coming to realize that trying to mold themselves to "fit" into an unaccepting Western culture is violent and destructive for the individual and the collective, and it will not solve Africa's problems. It was through these kinds of reflections and discussions that the Shembe movement was established as both a spiritual and passive resistance movement. And over a century later, it is also through these reflections that students challenged the violence of the state and universities during FMF. It was through the students' efforts and calls for radical and immediate transformation that these once lost ideas were brought to the forefront to be discussed, researched and implemented. Future African scholars, particularly those of color, have a responsibility and inherent duty to explore and build on this kind of research and the struggles FMF students raise, not just in the social sciences, but also in their various fields of sciences, commerce, mathematics and so on. This is the work that needs to be done to dispel the myth that Africans were not creating knowledge before Western occasion and to bring those knowledge systems (old and new) to the mainstream, both in the continent and all over the world. There are many challenges that will be faced when attempting to do this, some personal and internal, while others will be external and structural. The personal challenges speak to the fact that Africans have not healed as individuals or as communities. The centuries-old oppression is entrenched in many Africans' minds; they believe and sometimes perpetuate the idea that they are inferior. The traumas experienced decades or centuries ago are transferred from generation to generation. As ideological as it sounds, Africans need to go through individual and collective healing, a process of "decolonizing the mind". There are many ways to do this and these should also be researched and debated, but it is my opinion that one of the best ways we can do this is through reclaiming African spirituality as a cross-cutting guiding principle that informs different knowledge discourses, daily life, sustainability, social interaction, business, and so on. Many of the external and structural challenges that will be encountered have already been mentioned in part throughout this paper; they include challenges participating in the global stage where blackness is undermined, where the global north is skeptical of buying into ideas and strategies led by Africans, particularly black Africans. The historical and continued erasure of black or African contributions in meaningful advance and the appropriation of those advances by white people and the West at large is another major structural challenge, but it is through this kind of research that these systems will be challenged and dismantled.

CONCLUSION

This paper highlights many of the issues and challenges in South Africa's educational and knowledge production structures, especially those that have their origins rooted in Western or colonial history, which sees itself as superior to African history. This was done primarily by exploring South Africa's history of protesting and resistance within schools and universities. However, this paper in no way suggests that people boycott schools and mainstream educational or health structures, as they still have their importance and relevance, especially in a modernized space like South Africa. It would also be a mistake to completely reject these structures when not enough work has been done to build up African systems and ideas. The contradiction, however, is that while we need mainstream education for our survival, especially in the modern world, it is also rooted in oppression and creates an inferiority complex amongst Africans who are not represented in the structure. The intention of this paper is to dispel the myth that African ideas are inferior, whilst showing a glimpse of the history that led to this false sentiment. Like the #FMF student movement, this paper calls for decolonized social structures, especially in education. Our country cannot survive without this change, as racial and political tensions rise every day. Africans have a lot to teach the world, and empowering Africans to understand and be proud of their culture and the knowledge that comes from it, has the potential to fix many of Africa's and South Africa's problems, including high unemployment, challenges in accessing healthcare, a failing school system, poverty, environmental and technological challenges and many more. Taking steps to prioritize African languages and cultural systems within "mainstream" education, may also be the restorative tool needed to address the festering apartheid wounds and burdens still carried by many black South Africans. Moreover, giving more importance to African languages and ideas in schools and universities may save our school systems as knowledge will be delivered in a language in which more students are comfortable learning. Such content would also instil the missing notion that black people are enough and can be doctors, scientists, inventors, farmers, entrepreneurs, environmentalists, writers and more, without turning their backs on their culture, because their culture is the thing that can get them to where they need to go. This discussion includes looking at Helichrysum, a herb usually used in African cultures for religious reasons, but which also has important health and scientific purposes, including curing diseases and illnesses in a way that is significantly cheaper and not as risky as Western medicine. If Helichrysum alone has so much potential, what about the other African science and knowledge we ignore while we obsessively chase all that is Western in not only medical but also educational systems?

REFERENCES

Ani, M. (1994). *Yurugu: An African-centered Critique of European Cultural Thought and Behavior*. African Word Press.

Becken, H. J. (1968). On the Holy Mountain: A visit to the New Year's Festival of the Nazaretha Church on Mount Nhlangakazi, 14 January 1967. *Journal of Religion in Africa. Religion en Afrique*, *1*(2), 138–149. doi:10.2307/1594789

Bernasconi, R. (2001). With what must the philosophy of world history begin? On the racial basis of Hegel's eurocentrism. *Nineteenth-Century Contexts*, *22*(2), 171–201. doi:10.1080/08905490008583507

Biko, S. (1998). *The African Philosophy Reader* (P. P. H. Coetzee & A. P. J. Roux, Eds.). Routledge.

Biko, S. (2004). *I write what I like: a selection of his writings/ Steve Biko* (C. R. Aelred Stubbs, D. Tutu, & N. Biko, Eds.). Picador Africa.

Blaut, J. M. (1992). The theory of Cultural Racism. *A Radical. The Journal of Geography*, *23*, 289–299.

Bonetto, S. (2006). Race and Racism in Hegel: An Analysis. *Minerva – An Internet. The Journal of Philosophy*, *10*, 35–64.

Cabral, A. (1969). *Revolution in Guinea: An African People's Struggle*. Monthly Review Press.

Christie, P., & Collins, C. (1984). Apartheid Ideology and Labour Reproduction. In P. Kallaway (Ed.), *Apartheid and Education: The Education of Black South Africans*. Ravan Press.

Cock, M., Dold, T., & Wiersum, F. (2003). The Cost of Culture. *Veld & Flora.*, *89*(4), 146–149.

Dougall, J. W. C. (1963). *Christians in the African Revolution*. The Saint Andrew Press.

Douglas, H. P. (1979). The Passive Voice Theories in Religious Sociology. *Review of Religious Research*, *21*(1), 51–61. doi:10.2307/3510155

Fanon, F. (1964). *Towards the African Revolution*. Grove Press.

Fernandez, J. W. (1973). The precincts of the Prophet: A day with Johannes Galilee Shembe. *Journal of Religion in Africa. Religion en Afrique*, *5*(1), 32–53. doi:10.1163/157006673X00041

Griffin, S. F. (1995). 'Shembe' is the way. Durban: Institute of Current World Affairs (ICWA).

Hyslop, J. (1989). A Destruction Coming In Bantu Education as a Response to Social Crisis. African Studies Institute.

Jenks, C. (1993). *Culture*. Routledge.

Johns, S. (2004). Isaiah Shembe and the Uses of Textuality. *Journal of Southern African Studies*, *30*(2), 401–403.

Kumar, R. (2005). *Research Methodology: A Step-by-Step Guide for Beginners*. Sage Publications.

Lourens, A. C. U., Viljoen, A. M., & van Heerden, F. R. (2008). South African Helichrysum species: A review of the traditional uses, biological activity and phytochemistry. *Journal of Ethnopharmacology*, *119*(3), 630–652. doi:10.1016/j.jep.2008.06.011 PMID:18606217

Mason, J. (1994). Linking Qualitative and Quantitative data analysis. In A. Bryman & R. G. Burgess (Eds.), *Analyzing Qualitative Data*. Routledge. doi:10.4324/9780203413081_chapter_5

McCulloch, J. (1983). *Black Soul, White Artifact. Fanon's clinical psychology and social theory*. Cambridge University Press. doi:10.1017/CBO9780511558559

Meersman, B. (2012). Soaring search for truth. *Mail and Guardian*. Retrieved: September, 2012, from www.mg.co.za/print/2012-05-04-soaring-search-for-truth

Mndende, N. (2009). Tears of the Distress: Voices of a Denied Spirituality in a Democratic South Africa. Dutywa: Icamagu Institute.

Mpanza, M. P. (1999). *Submission to the Truth and Reconciliation Commission by Ibandla Lama Nazareth*. Retrieved: June, 2011, from web.uct.ac.za/depts/ricsa/commiss/trc/nazar.htm

Muller, C. A., & Buchanan, D. A. (1994). *Nazarites song, dance, and dreams: the sacralization of time, space and the female body in South Africa*. New York: UMI Dissertation Services.

Ndhala, A. R., & Staden, J. V. (2012). Smokescreens and mirrors in safety and quality of herbal medicines: A case of commercialized herbal preparations. *South African Journal of Botany*, *82*, 4–10. doi:10.1016/j.sajb.2012.06.008

Ndlovu, S. M. (2006). *The Road to Democracy in South Africa*. Academic Press.

Ndlovu-Gatsheni, S. J. (2013). *Coloniality of Power in Postcolonial Africa Myths of Decolonisation*. CODESRIA.

Nkrumah, K. (1965). *Neo-Colonialism, the Last Stage of Imperialism*. Thomas Nelson and Sons.

Pauw, B. A. (1963). African Christians and Their Ancestors. In V. E. W. Hayward (Ed.), *African Independent Churches Movements*. Edinburgh House Press.

Posnock, R. (1997). How It Feels to Be a Problem: Du Bois, Fanon and the 'Impossible life of the Black Intellectual'. *Critical Inquiry*, *23*(2), 323–349. doi:10.1086/448831

Sogiba, Z. S. (1996). *Steve Bantu Biko: Politician, 'Historian' and 'Proponent' of African Traditional Religion*. University of Cape Town.

Tarsons, P. (1958). *Max Weber: The Protestant Ethic and the Spirit of Capitalism*. Charles Scribner's Sons.

Union of South Africa. (1953). *Bantu Education Act, Act No 47 of 1953*. Retrieved: August, 2017, from http://www.sahistory.org.za/archive/bantu-education-act%2C-act-no-47-of-1953

Zagzebski, L. (2004). Foreknowledge and Free Will. *Stanford Encyclopedia of Philosophy*. Retrieved: June, 2012, from: https://plato.stanford.edu/entries/free-willforeknowledge

Zondi, N., & Ntshangase, M. (2013). Shifting Boundaries in Respect of Perceptions Governing Women and the Burning of *Impepho. Alternation (Durban), 20*(2), 213–235.

KEY TERMS AND DEFINITIONS

African Independent Church(es): Churches that subscribe to African Traditional Religious ideals. Some are big, organised and unite different cultural groups in the way that the Shembe church is, but few are as large or diverse. Many are formed across common cultural identity lines, which is why there are so many different AICs; the different cultures that bind them might have similarities at times, but they are typically vast and diverse.

African Traditional Religion: Should preferably be referred to as African Traditional Religion(s) because there are many and they do not carry similar characteristics in the way that Ibramic faiths do. In fact, some reject Ibramic faiths, while others embrace certain aspects and Africanize them. One of the few things that most ATRs have in common is that they respect and honour ancestors.

*Amagqobheka:***:** Black Christian converts, forbidden to use African herbs and engage in African practices or rituals.

*Amaqaba:***:** A derogatory word often used to describe uneducated black people. It loosely translates to "primitive", "backwards" or "pagan".

Bantu Education Act: The act of 1953, also known as the Black Education Act. It led to the erasure of African indigenous languages in schools. Black students were forced to learn in Afrikaans and were only taught low labour skills that would aid in assisting whites.

eBuhleni: Shembe Village located in Inanda, KwaZulu-Natal. It is a Shembe "mecca" and a sacred and holy space. It is where gatherings take place in July and September each year. The direct English translation is "a place of beauty". eBuhleni is the biggest Shembe faction.

ekuPhakameni: Shembe Village located in Inanda, KwaZulu-Natal, It is a Shembe "mecca" and a sacred and holy space. It is where gatherings take place in July and September each year. The direct English translation is "a higher place of enlightenment". This branch is the oldest Shembe faction.

Fees Must Fall (FMF): South African national student movement, characterized by protests across different universities. It challenged structural and systematic exclusions of marginalized students and employees in university spaces, calling even for a decolonized curriculum. The movement peaked in 2015 and 2016, but arguably still exists today.

Impepho:: African herb/plant burnt when awakening/communicating with the ancestors. It can also be used for medicinal purposes. The scientific name of the genus to which it belongs is Helichrysum.

Inanda: Predominantly black township located in KwaZulu-Natal, South Africa.

Isaiah Shembe: Founder of the Shembe Church.

Johannes Shembe: Isaiah Shembe's third wife's son. He took over the church after Isaiah's death in 1935.

Mokgoro:: Sesotho name for a traditional mud house.

Mount iNhlangakazi: Sacred Shembe site where the annual pilgrimage takes place each January.

Orania: Located in Northern Cape, South Africa. It is a whites-only Afrikaner town, where black people are typically only allowed in as low skilled labour.

Sangoma: Diviner, spiritual healer, someone who can connect the living with their ancestors through rituals and prayer.

Shembe Church: An African Traditional Religious church that was founded by Isaiah Shembe in 1910. The church mixes African tradition and culture with what it believes to be the best parts of Christianity.

Soweto Uprising: A series of demonstrations and protests which took place in 1976, whereby students challenged the Bantu Education Act, as well as apartheid as a whole.

Chapter 2
The Ambit of Ethics in the South African Academic Institutions:
Experience of Coloniality

Xolani Mathews Shange
University of South Africa, South Africa

ABSTRACT

The chapter examines possible use of instruments and processes such as ethical clearance in the institutions of higher learning as subtle means of perpetuating inequality and racial prejudice towards the indigenous people of South Africa who had recently emerged from the scourge of apartheid with a hope of democracy ultimately providing not only freedom of association and speech, but also intellectual freedom. Freedom to produce African-based knowledge by Black African intellectuals pursuing their postgraduate studies and academics whose careers are at formative stages. However, their vision of becoming producers of African Indigenous knowledge is thwarted by subtle and invisible activities that are aimed at perpetuate coloniality in the higher institutions of learning. Sadly, ethical clearance process has possibly been utilized to derail research outputs that some of the old guard from historically white universities are uncomfortable to witness, thus continuing to maintain the colonial status quo.

DOI: 10.4018/978-1-7998-1249-4.ch002

INTRODUCTION

South Africa was the last country to obtain independence in Africa and the universities underwent changes through the colonial, apartheid and the democratic era. In each period, they had found themselves influenced by the political system of the moment. In particular, colonising forces in the past centuries utilised the South African system of education to engrain a narrow cultural perspective (Beckmann, 2016; Do Vale, 2016, p.600) and in the process propagating racial inequality especially in universities. The existence of colonialism and apartheid became a root course of alienating black African students and academics from academia. The situation has been overturned to consider approaches that include rather than excludes various inhabitants of South Africa, especially in accessibility and academic freedom. There was also recognition that such participation should not only include black students and women but to also guarantee successful outcomes (Akoojee & Nkomo, 2007, p.386). However, there seem to be continuous lack of political will to transform not only the curriculum of the universities, but also the inclusion of African ethics in the educational land scape. Collins and Millard (2013) contend that it is unwarranted for the historical disadvantaged communities to be assimilated into a system that overlooks their cultural values. Therefore, there should be transformation of the education system in order to entrench relevance to the goals of a democratiç South Africa (Pinar, 2009, p.318). Although promotion of education for all has been pronounced, the higher education system continues to be engulfed with black African student higher dropout (Strydom, Kuh & Mentz, 2009, p.260) and failure rate coupled with ethical and moral practices as well as standards that are not embracing African values. Meanwhile African governments and institutions have lost their morality and are torn apart by structured inequality and are deeply polarized along ideological, religious, racial, gender and ethnic lines (Suda, 1996, p.72). The situation warrants for stakeholders in education and government institutions to deeply reflect on concrete strategies to entrench sound and liberatory education system for all inhabitants (Beckmann, 2016, p.289; Do Vale, 2016, p.593).

This chapter postulates that ethical research aspects at South African universities derive their practices from the colonial Eurocentric philosophy which continue to co-exist through coloniality and disregard for the African context, norms, values and practices. As Cloete & Auriacombe (2019) concur that "*colonial values and mental models were introduced in educational institutions and processes, in most cases as formal substitutes for pre-colonial values and mental models*". Furthermore, the African countries' research ethics standards have been largely influenced by the internationalisation of research through colonialist history (Ndebele et al, 2014, p.4). In further analysing this matter, an uBuntu framework will be highlighted to reflect on the limitations of Western ethical theory and methods while highlighting strengths of the indigenous knowledge related ethical research aspects (Peltier, 2018, p.2) so as to critically analyse ethical research issues in the South African universities.

BACKGROUND

Apartheid policy reigned in the South African education system, with a skewed curriculum that accommodated only white minority cultural values (Kallaway, 2002, p.2). Blacks were furnished with inferior education to ensure they remained unskilled and semi-skilled workforce (Kallaway, 2002, p.12; Hale, 2010, p.8) of which at a later stage they would uplift the white capitalist economy through provision of hard labour. The few black people who made it to the university level could be admitted at historically white university colleges while they required permission from the minister to be enrolled at the historically white universities (Beale, 1998, p.2). However, the universities were under close watch of the repressive regime such that the right to academic freedom was suppressed not only for the black African scholars and students, but for all races. In 1700s there was concerted effort by the Dutch authorities in South Africa to prevent perceived subversive publication to be published (Delmas, 2011, p.116). The British suspended publications of material considered as of political nature a century later (Oliphant, 2000, p.11).

The apartheid government intervened in universities to counter the emergence of political resistance to the apartheid policy and therefore invoked censorship legislation. As a result, several universities were controlled by the Broederbond leading to government intervention in preventing employment of some black academics. For instance, in, 1968 the University of Cape Town prevented appointment in the Department of Anthropology, scholar, Archie Mafeje, through the intervention of the state (Leruex, 2013, p.129). Colonial rule was deeply entrenched and the ethical morals were at their lowest.

Over the past two decades, there has been some transformation in the higher institutions of learning landscape, in particular on the curriculum with introduction of multicultural educational approach (Garriott, Reiter & Brownfield, 2016, p.158). However, the slow rate in incorporating African ethical values and norms in the research front continue to be cause for concern. The conventional educational approaches as it pertains to assessment and teaching continue to perpetuate poverty to the previously disadvantaged individuals (Connell, 1993, p.24). Beckmann (2016, p.299) argues for a better comprehension of such ill-defined debates as consisting cultures, attitudes, and styles that impede the goal of transformation that embrace different cultural backgrounds. For ethical research consideration approaches to embrace African context, there should be transformation of learning environments and institutional cultures in the education system, given the continued prevalence of upsetting and exclusionary elements that are keenly maintaining colonial status quo. The inclusion of multicultural values within the higher institution environment should be prioritised and its delayed installation will from time to time ignite anger to the inhabitants of the country, especially the youth.

This chapter maintains that in order to effectively deal with the inequity, strategy and policy formulation ought to acknowledge predicaments and limitations pertaining to transformation and for ethical research consideration approaches to embrace African context, there should be transformation of learning environments and institutional cultures in the education system, given the continued prevalence of upsetting and exclusionary elements that are keenly maintaining colonial status quo. In view of plethora of responsibilities and challenges, learning and teaching methods should be lined up with contextual scholarships, realities, and conceptualisations (Higher Education South Africa, 2014, p.4). Research as conducted in the higher institutions of education is aimed at contributing towards development and production of new knowledge. Such knowledge should be of value to the community within which it is produced such that it is best positioned to solve immediate problems of those communities, especially those of the disadvantaged groups. Furthermore, it should develop sufficiently rigorous local knowledge that relates better to the needs of students and the development challenges of South Africa, while contributing to global knowledge production from the perspective of Africa (Council on Higher Education - CHE, 2017, p.4). Nonetheless the current higher education setting is biased towards a Westernised and Eurocentric ethical research trajectory while less of the African cultural values are embedded in it. This argument flows from the reality that "*pre-colonial knowledge and value systems that were contradictory to the colonial way of life were in most cases disrupted, abolished or changed to comply with the new status quo*" (Cloete & Auriacombe, 2019). Therefore, the kind of solutions derived from the current and prevalent research are rooted in the past colonial culture thus perpetuate un-Africanness and un-uBuntu behaviour that is contradictory to the Africanist ancestral teachings and spirit. Hence Africanisation should focus on changing education and culture (Wa Thiong'o, 1981, p.88).

The colonization impact on education led to modification of social values from natural abilities and lifelong learning values to those that are aligned to labour market (Do Vale, 2016, p.595; Mthembu, 2017, p.185). Therefore, when the community is faced with challenges and need to overcome them, the solutions offered by the Eurocentric research values often are unsustainable and individuals who graduates from such educational system are unable to be fully self-reliant and creative, thus reinforces capitalistic and Eurocentric thought. It is against these reasons that inter-generational poverty persists in previously disadvantaged communities based on societal inequality and high rate of unemployment.

These inadequacies in compensatory higher institutions of learning may contribute to higher dropout rate thus contributing to continued vulnerability and unemployability. Amongst the reasons often cited for the increased unemployment rate in South Africa is the colonial history and structural economic challenges. Structural unemployment is often viewed as responsible for misalignment between the surplus labour and skills endowment of individuals, which is an additional labour force that is always available for the effective operation of the capitalist system (Mlatsheni, 2012, p.32). Apartheid was instrumental in the oversupply of black unskilled and semi-skilled labour to aid the growth of capitalist enterprises that operated mainly in the mining and agricultural sectors. South Africa has witnessed relegation of previously disadvantaged individuals to second class citisens through being affected by the scourge of high unemployment. This is accompanied by usurping their right to acquire relevant economic skills that would have been instrumental in securing jobs in the modern and sophisticated economic landscape.

The argument presented in this chapter is that higher education research contributes little in providing practical solutions such as eradication of unemployment in the most vulnerable communities. I further imply that those who do not finish their research related studies due to stringent Eurocentric research ethical considerations are likely to be unemployed or become employed in very junior positions and earn slavery wages. There are no guarantees of employment for those who happen to finish their research studies and obtain their qualifications. It could be argued that Eurocentric research practices perpetuate poverty, inequality and unemployment.

The type of higher education and research skills acquired should influence employability and self-reliance of individuals coming from diverse cultural backgrounds so that in turn are able to solve their communities' immediate challenges. The African based research ethical focus should promote accessibility to higher degree research by the previously disadvantaged individuals, support increased pass rate accompanied by the quality of African culturally embedded research that helps to eradicate poverty, unemployment and inequality in society. Furthermore, such research should support incumbents with appropriate skills to enhance their abilities to become self-reliant, not to swell the ranks of the unemployed.

It is apparent that African countries including their higher institutions of learning have either remodelled or established their research oversight committees and systems on World Health Organisation (WHO) or the United State institutional review boards system that provided guidelines on the procedures of research ethics committees (Ndebele et al, 2014, p.4), yet the very institutional review boards have been highly criticised. Meanwhile efforts had been thwarted to instil a truly African university that draws motivation from Africa, and intelligently devoted to her aspirations and ideas, modelled from the seed growing from African soil (Ndlovu-Gatsheni, 2017, p.12).

In summary, issues of research ethics that are African inclined, their influence in building quality research capability within the previously disadvantaged groups to bolster their self-reliance and ability of such research to solve immediate community challenges depend on a truly transformed and decolonised African university. However, there are several risks associated with universities' decreasing value of scholarship linked to spending more time to publish as much as possible rather than researching about what really matters to drive economic and social agenda (Trust, 2019).

MAIN FOCUS OF THIS CHAPTER

This chapter will outline shortfalls of the current research ethics in universities emanating from the inability to embrace multiculturalism in that the African ethos, norms and values are incorporated. This argument will be accompanied with a highlight of African morality and its inclination towards supporting the African production of knowledge perspective. The ethics theorists state that universities operate as transmitters of predominant values (Pinar, 2009, p.318). So, the increased demand for a multicultural system of education necessitates a high level of respect and comprehension from individuals belonging in various cultures (Cajetas-Saranza, 2015, p.39).

Meanwhile ethics are seen as moral judgements, values and norms that guide behaviour which includes morality that clarifies what is regarded as correct, not to do harm, good and right (Schenck, Nel & Louw, 2015, p.99). What is moral about not considering the majority people's culture whose roots are embedded in the soil where academic and research institutions stand and operate? Whose interests are served with this kind of hypocrisy?

There is a need to accommodate a multiple cultural perspective when considering policies and strategies that guide ethical research committee work at the universities. In retrospect, Canterbury Christ Church University (CCCU) (2014, p.3) argue that ethical issues are based on respect for human dignity, truth (veracity), autonomy (self-determination) and justice.

Thus the research ethics should also encompass African perspectives including the African philosophies of uBuntu, acknowledgement of the role played by a community and women in instilling morality. Acceptable ethics should find expression in the research ethical consideration at institutions of higher learning. The chapter put emphasis on cutting umbilical cord of apartheid and coloniality legacy in institutions of higher learning through supporting research ethics that are pro-African, reflecting the content and context of indigenous African people of South Africa.

CONCEPTUAL FRAMEWORK

For the purpose of this chapter, coloniality, is perceived as the impact of colonialism in the formerly colonised territories that persists many years after physical colonialism has subsided. It points to the continuity of colonial forms of power and domination after the end of direct colonial administrations (Grosfugel, 2007). As Dastile & Ndlovu-Gatsheni write, that coloniality of knowledge relates to direct epistemological colonisation whereby Euro-American techno-scientific knowledge accomplished destroying, displacing, disciplining alternative knowledge it found outside the Euro-American zones while appropriating what was considered valuable to the designs of global imperial (2013, p.110).

Colonial curriculum divided the African continent in terms of spirituality, ethnically, economically, socially, physically, psychologically and culturally as alluded in chapter 3 of this edition titled '*A deconial curriculum is everything: An Afrocentric approach*' in p8. Hence more efforts are needed to dismantle this episteme colonial effect. However, much of South African writing on curriculum decolonisation appears to be based on the Latin American branch of decoloniality theory (e.g. Mignolo, Quijano, Maldonado-Torres, Grosfugel) which has roots in Marxist dependency theory. Both streams build on the work of Fanon which puts at the centre of the 'decolonial epistemic perspective' the:

"multi-faceted struggles over subjectivity and negative representations, over the imposition of Euro-American epistemologies, and over exploitation and dispossession. Decolonial epistemic perspective's mission is to forge new categories of thought, construction of new subjectivities and creation of new modes of being and becoming" (Dastile & Ndlovu-Gatsheni, 2013).

uBuntu refers to high moral ground as envisaged by black African ancestors who engrained this seed in all aspects of their lives as they respected human beings whom they regard as equals and at all times kept the motto of 'do unto others as you would like them to do unto you': 'A human being sustains her livelihood because other human beings exist to support her and vis a versa'.

Indigenous African people is used interchangeably with black people. Black people are those sons and daughters of the South African soil whose ancestors are of African or Ethiopian continental origin, suffered brutality of African version of colonialism. African culture is referred to various black South African way of life that is historically and currently practiced in the four corners of South Africa, including its variations.

The author utilised systematic observations based on practical first-hand experience as well as secondary literature analysis in writing this chapter.

NOVELTY OF MORAL STANDARDS WITHIN WESTERN ETHICAL PERSPECTIVE

The history of ethics in research dates back to Nazi atrocities during the Second World War which, significantly violated the basic principles of human rights as a result of which the first code of practice for ethical research was developed [Dempsey and Dempsey, 1992 (The Nuremburg Code of Ethical Practice); CCCU, 2014, p.2)]. Nazi atrocities alongside with Tuskegee experiment are mostly cited examples where researchers blatantly disregarded the dignity and well-being of research subject (De Wet, 2010, p.302). The ethical research requirements laid down by the institutions of higher learning through the Research Ethics Committee (REC) emanated from the medical research where immediate danger could be posed during experimentation with medications and procedures (Zimmerman, 2005, p.304). For instance, during the Tuskegee Syphilis Study from 1932 to 1972 there were large number of black males who were utilised in the clinical trials and became extremely negatively affected (Welsing, 1991, p.iv). For social research, REC approaches are often misty and demands made on scholars seem difficult because the effect of their activities seem largely less clear (Zimmerman, 2005, p.304).

The Declaration of Helsinki (Appendix 1) illustrates the ethical issues that must be considered in undertaking research involving human participants. Such issues are based on respect for human dignity, autonomy (self-determination), truth (veracity) and justice (CCCU, 2014, p 3).

There was development of the recommendations involving human subjects (Declaration of Helsinki) adopted by the World Medical Assembly in 1964 and most recently updated in 2013 (CCCU, 2014, p.2). This was put in place especially in social sciences to deal with the risk associated with psychological harm, unfulfilled expectations, deception, unexpected or erroneous representations and different interpretations (De Wet, 2010, p.302). The implication being that harm could also manifest itself at post research stages. Therefore, research ethics also mean respecting participants, minimising harm, ensuring confidentiality to the greatest degree, and engaging in meaningful and authentically informed consent (De Wet, 2010, p.303).

The wave of neoliberalism spread from Europe to Africa in the mid-90s led to a call for South African universities to be managed as privately owned companies. These attempts were infect intended to commodify knowledge and silence any form of political dissent in Universities. Also RECs were considered as a certain form of 'censorship' or an "*academic freedom infringement*" (Kim et al. 2009, p.535).

With the demand to expect universities to increase their research output, there was enhanced application for research funding from institutions external to the universities. Some cohorts in the academia saw this as an opportunity to make profit through research, adding salt to injury to the earlier emergence of extreme inhuman and brutal handling of research especially when human subjects were involved. Based on these developments the Eurocentric oriented research ethics as now administered by the ethical research processes were established to ensure that human subjects were not abused. However, this process has not incorporated the perspectives of the African cultural context and multiculturalism and is without uBuntu as a foundation to this process such that developments could easily be abused and used as a tool to perpetuate coloniality in the higher institutions of learning.

The research ethics should incorporate uBuntu seen as a religious way of life in African societies' pre-colonial period where it was amplified that each person lives for the other and therefore sacrificing for others, protecting and caring in a selfless way for other human beings. It also shows that everything in my environment is crucial for my existence as well as that of fellow human beings (Msila, 2015, p.1). It is un-uBuntu if the ethical research practices consider certain aspects of the environment while ignoring others.

The Western research justified the development of policies to guide every aspect of indigenous peoples' living (Tuhiwai -Smith, 2012, p.3). It was questioned at a lesser extent the Western research that told the indigenous people:

"things already known, suggested things that would not work, and made careers for people who already had jobs" (Tuhiwai -Smith, 2012, p.3).

It is such research ethics procedures that were criticised as having legal jargon and yet expected to be understood and signed by the local rural communities for consenting into conducting of research.

SOUTH AFRICAN UNIVERSITIES AND ETHICS

South Africa in particular experienced two phases of coloniality project aimed at side-lining the African indigenous people. Firstly, there was colonization by the English people and later by the Afrikaners who ensured introduction of apartheid and explicitly introduced policies that excluded black people from the mainstream production of knowledge especially at the higher institutions of learning. During this historical epoch, apartheid government appointed their individuals in all black universities who were fresh graduates from the Afrikaans medium universities and were quickly promoted (Le roux, 2013, p.129), to ensure that they sort of served as guards and applied punitive measures to those black academics who were getting out of hand (Le roux, 2013, p.130).

Academic freedom for the Afrikaners was narrower in that it meant protection of their culture and language within its academies (Le roux, 2013, p.131). In the establishment of universities there was a 'conscience clause' meant to protect students and staff from discrimination on the basis of their opinions and beliefs, but this clause was only meant to protect religious views rather than the political ones (Le roux, 2013, p.131). However, there were no such clauses in the establishment of historical black universities which ostensibly undermined their status of being properly constituted institution of higher learning (Le roux, 2013, p.131).

Between 1960s and 1970s restrictions were imposed to academic freedom at historical black universities. For instance, some rectors of university colleges were empowered to control student relations to the press and publication (Le roux, 2013, p.133). Universities such as University Cape Town (UCT) and Wits spoke aggressively against suffocation of academic freedom while other staff and students were not agreeing. This culminated with the publication of booklets titled '*The open universities in South Africa*' where they wrote definition of academic freedom (Le roux, 2013, p.133).

An intellectual should be independent, rebel against hidden and open pressure and manipulations, should be the chief doubter of systems, of power and its incantations (Karabel, 1996, p.205). We doubt the intent of some individuals in the colonial educations institutions to remain truthful to this oath hence some never spoke against atrocities of apartheid. Even the intellectuals and academics at Wits and UCT who protested against academic freedom by publishing a booklet where they defined what it meant, they avoided to criticise the apartheid regime.

This happened partly because they were drawing privileges from the white power structure in the form of the repressive apartheid regime and could not want to lose out. How are we supposed to celebrate the bravery of these white liberal academics when their actions proved to be careful not to provoke the apartheid government? We remain sceptical about their intentions, except few white liberal academics who were eventually forced into exile and others jailed and killed for speaking the truth against the censorship of academic freedom and apartheid oppression of black African people in general. These events happened in addition to the earlier displacement of the indigenous African systems of education during the establishment of the modern university in Africa (Ndlovu-Gatsheni).

When the colonial grip was slowly loosening up, fewer indigenous African black people were allowed to enter into higher institutions of learning under stricter conditions. For instance, a black person was expected to obtain permission from the Minister of Education to be allowed to study at the former white universities. This also meant interference by apartheid government and lack of academic freedom and institutional autonomy (Kori, 2016, p.49).

The repelling of black African race from higher institutions of learning began directly through formal censorship policies at universities especially in the open universities derived from the relevant country's segregation legislations. White race dominated these open universities and underwent evolution until in the post-apartheid period where black led government enacted legislations that allowed intervention by the minister of education to fastrack transformation. The historical white institutions increased admission of black African people but indirectly continued to censor intellectual output by the African students and academics using various means. A case in point is an outcry by Kiguwa (2019), articulating her experience regarding subtle exclusionary cultural practices in academy saying that there are:

"almost invisible nature of informal institutional cultures within departments; these function to reinforce and sustain hegemonic practices, spaces and traditions of the academy that are exclusive to already marginalised bodies. These cultures and communities of practice are so invisible as to render any challenge and resistance almost impossible, precisely because one would first have to undertake the work of making the invisible visible......How do black bodies navigate spaces that, on the surface, seem to be open and yet may be functionally immutable because of this openness?" (Kiguwa, 2019, p.11).

This statement by a black academic in the former white academic institution reveals stack of evidence that black academics and students were forcibly allowed to be party of these institutions through repealed apartheid laws and therefore democratic compliance had to be embraced. The committees or communities of practice are racially mixed and on one hand some black academics had been co-opted into these invisible white academic cultures in a stealth manner and they may not be aware while on the other hand there are black academics who are conscious about their surroundings as well as self-conscious about their actions and reaction in relation to these situations. Hence black female academics working in such institutions were brave and wrote what they felt about the succinct exclusion of blacks and particularly females in the institution of higher learning in their book titled '*black academic voices: The South African Experience*' (2019). More often than not, it is not easy to uncover these invisible behaviours and cultures, and therefore difficult to fight back. When faced with such situations, highly black conscious academics formed networks with other black academics and support staff in various departments and colleges to garner support in order to execute a coordinated black academic transformative and decolonisation agenda as well as taking steps to negate coloniality in institutions of higher learning. This is done with a view to influence society so that black people are counted in these academic spaces despite prevalence of exclusionary institutional cultures where non-whites are expected to adapt instead of promoting change and inclusive culture (Khunou et al, 2019, p.7). These subtle discriminating academic institutions maintain status quo of rigid white culture and further establish double standards where white and black academics are treated differently in terms of promotions and access (Khunou et al, 2019, p.8). Moreover, this creates negative perceptions about what black academics can contribute, hence Khunou (2019) assert that "*being different in academia succinctly captures how marginalisation and silencing tactics are used to exclude black migrant academics and students from the academy*" (Khunou et al, 2019, p.8).

These retrogressive practices in academia derail grooming of black academic change agents that should champion further academic transformation, leading to broad economic and social emancipation of black communities in South Africa.

Some black academics have seen a need for "*problematisation of blackness and whiteness as significant for the transformation of the academy*" (Khunou et al, 2019, p.8). More questions have been posed such as how struggles emanating from one's history of education can potentially facilitate decolonial ways of teaching (Khunou et al, 2019, p.9). It is suspected that this line of thinking poses a threat in the former coloniser who fears replacement of white by black intellectual domination in academia, hence there is greater blockading of black Africans in the academic spaces.

This chapter therefore argues that various coloniality attuned tools and techniques such as ethical clearance existed in higher institutions of learning during the colonial era and continued in subtle and invisible ways during the post-apartheid dispensation. Such techniques include operations of RECs, ethical clearance processes and procedures. As a result, fewer research based post graduate outputs amongst black African people were produced especially from the historical open universities. Most of the hindrances existed in the research ethics committees, delays in issuing the clearances leading to black African students quitting their studies before they begin with fieldwork. What exists now is an 'unethical clearance of the soul' by few white racist and co-opted black academics who continue to be the agents of coloniality to subtle suppress the black African intellectual voice.

On one hand Steve Biko spoke of how colonialism led to imposition of self-hate by the black African people because they had internalised propaganda to believe that they were worthless, useless and inferior, hence the 'pathological fear 'phenomenon (Pityana, 2008, p.5) that engulfed black people became detrimental and had to be eradicated. On the other hand, majority of white bodies during the colonial era internalised the propaganda to believe they were a superior race, powerful, the best and therefore had internalised what I term 'racist based self-actualisation'. Racist based self-actualisation could be described as an ultimate state of consciousness by a racist where he/she feels and believes he/she is above the 'other' race perceived as subservient, leading to an extent of disregarding the existence of such race, have a sense of belief that the other race will never make any positive contribution to betterment of socio-economic, scientific and technological advancement of society. Other scholars such as Bourdieu & Passeron 1990, p.56 also refer to a notion of 'habitus' that relates to 'socialised subjectivity' and 'embodied constitution'

(Kiguwa, 2019, p.15). This embodied institution is associated with a body that is "*literally moulded into certain forms, so that the habitus is reflected in the whole way that one carries oneself in the world, the way that one walks, speaks, acts, eats*" (Thompson, 1984: 54; Kiguwa, 2019, p.15). Therefore 'white habitus' as a "*socialisation process is fundamental to understanding how white bodies justify and maintain practices and attitudes of difference and segregation*" which is also seen as sustaining "*exclusionary practices by white bodies of black staff members within higher learning institutions*" (Bonilla-Silva, 2006; Kiguwa, 2019, p.15).

The 'racist based self-actualisation' and its association with deep internalisation of racism has continued to exist beyond colonialism and apartheid and have in some instances reached a stage when the racist believed he is above the law and lives of 'other' races are perceived inferior and unimportant thus suitable to be executed. Though this is an extreme state of racism where it has been witnessed especially being practised by some Afrikaner farmers who killed black people because they had mistaken them with animals (Bevan, 2006.), not all the time it reaches this stage. Most of the time the 'racist based self-actualization' has continued to exist in the households of racists where they feed racially based propaganda to their off-springs and extended families and to their unconscious African black workers. These racists tendencies propel racists to be nostalgic of racial segregation and nurture exclusivity of their racial group through use of monetary means. For instance, once a racist coming from a well to do background realises that black African people are buying property in his/her neighbourhood, he/she will opt to sell and leave the area in order to buy property elsewhere, most likely in upmarket neighbourhoods. However, in the institutions of higher learning this phenomenon is translated only to the level of exclusion of other race guarded by entitlement belief that they have ownership of the university systems as laid down and entrenched by their forefathers. Hence there is continued maintenance of hegemonic status quo in the knowledge, teaching, learning and research at universities (Heleta, 2016).

These racists have a clandestine network of like-minded individuals where they discuss about safeguarding their interests in a disguise of protecting minority rights. They turn a blind eye on the South African transformation agenda that ought to bring positive changes in academic institutions. They still believe that they could prevent black African students and academics from writing what they intellectually desire (Le roux, 2013, p.133).

These racists are nostalgic of the censorship tendencies that took place during the apartheid era that eliminated research papers that were unpleasant to the apartheid regime and to the racist bodies in general. These racist bodies would use any legitimate tool to subtle suppress the ideas and intellectual voice of the black African people, especially in institutions of higher learning.

I argue that research ethics may be utilised as subtle tool to derail and discourage black African students and academics from venturing into territories that have never been intellectually traversed. We speak of the 'ethical clearance of the soul', where it now depends on each committee individual rather than institutions to developing and applying fair research ethics and ethical clearance rules. It is against these reasons that the community must device means of holding institutions accountable and maintain intellectual struggle until epistemic and Eurocentric violence are dismantled at the South African universities (Heleta, 2016).

There is no obvious trace of ill-intentioned objective to undermine the black African intellectual voice through the design, use and control of the research ethical processes in these institutions due to subtle censorship. Also, the historical background to the formation of the universities of South Africa shows great entrenchment of the Eurocentric influence that has lived through the centuries with the undertone to sustain and deepen colonialism.

Therefore, institutions of higher learning have continued to embrace colonial culture in universities until in 2015 when students began to challenge its dominance. It was during this time that students called for de-colonilisation of the curriculum and other values and norms that continue to persist, giving a wrong impression that the universities still belonged to a certain segment of society that once unleashed colonial and apartheid rule in South Africa.

It is known that during the post-colonial period there were also calls for decolonisation which was seen at the same light as Africanisation. Such nation-building decolonisation call became subject to critique by intellectuals such as Fanon as he regarded such a project as aimed at benefiting the African postcolonial middle class. According to Fanon such benefit was associated with assimilation of black middle class into colonialist thought and therefore distorted Africanisation campaign to mean transferring those unfair advantages emanating from the colonial past legacy into native hands leading to ideological flaws reflected through chauvinism and xenophobia (Mbembe, 2016, p.33-34).

So the chapter argues that such colonisation continue to exist in various forms not only exclusive to research ethical process in the South African universities. Although apartheid laws had been physically removed from universities' statutes that are published in the university websites, the strife continues silently and invisibly.

COLONIALITY AND ETHICAL CLEARANCE

Coloniality is a global power structure that refers to long-standing patterns of power that emerged resulting from colonialism. These patterns also define knowledge production, labour, culture and inter-subjective relations. Coloniality is kept alive in books, in cultural patterns, in the criteria for academic performance, in self-image of peoples, in common sense, in aspirations of self, and various aspects of modern society (Ndlovu Gatsheni, 2018, p.24). It is difficult for the colonies to enter into higher echelons of global power and rather settled for lower ranks of the hierarchy where they are routinely spoon fed rules and norms by the 'big brother' from the industrial North (Gatsheni, 2013, p.3). South Africa is one of the world's countries that continue to suffer at the hands of global super powers and has widely opened her shores in vein with a hope of attracting investment to support local economic development and employment creation.

Emancipatory projects continue to show numerous weaknesses amongst which is lack of questioning Western modernity logic that enabled globalisation of Euro-American views (Gatsheni, 2013, p.4). This trajectory led into the construction of hierarchical, racialized, patriarchal, hegemonic and capitalist global social system (Gatsheni, 2013, p.4). The continuous and relentless intellectual assault of African and indigenous knowledge by the Western racist, patriarchal and dominant classes has rendered the African continent at the foot of global hierarchy.

Euro-American hegemonic knowledge banished alternative epistemologies from Africa and other parts of the global South to the barbarian margins of society and out of the global intellectual space (Gatsheni, 2013, p.4).

The dominance of Euro-Americans in knowledge creation has aided perpetuation of deliberate alteration of African history, as well as the cultural norms and values due to narratives that are biased towards the global North while causing havoc in the global South. Rodney (1972) alludes to the distortion of the African history pertaining to the dominance of the family principles in the "*communal phase of African development*" which affected largely labour, land and system of distributing goods (Rodney, 1972, p.37). Furthermore, he accuses the European anthropological researchers of allowing to be influenced by racism and prejudices in conducting research (Rodney, 1972: 37).

The coloniality project in the global South was aimed at plundering the colonies' economic and human capital resources which inevitably had to be accompanied by the colonisation of the mind and the intellectual resources. Colonialism went beyond conquest, occupation, annexation, settlement, exploitation and domination (Ndlovu-Gatsheni, p.181). It further consisted of emptying the brain of the native of the entire content as well as mutilating, altering, and ultimately destroying the history of the colonised (Ndlovu-Gatsheni, p.181; Fenon, 1968). The process of colonialism necessitated the introduction of strategies by the colonial masters to control the production of knowledge and sharing of such amongst the indigenous people of the African continent. This was important for colonialism because the oppressed people should not have made the intellectual growth strides given the potential of awakening the critical thinking that would form a critical mass to intensify resistance against coloniality project. The ethical clearances in the higher institutions of learning have also developed during the colonial era and it is not known the extent to which they have been utilized to prevent the emergence of the African indigenous intellectual voice. It may well have become like a new and subtle colonialism likened with the power of global coloniality that still suffocates the development and advancement of Africa today.

The notion that the elimination of colonial administrations amounted to the decolonization of the world equates to a 'postcolonial' world myth. The heterogeneous and multiple global structures put in place over a period of 450 years did not evaporate with the juridical-political decolonization of the periphery over the past 50 years. We continue to live under the same 'colonial power matrix.' With juridical political decolonization we moved from a period of 'global colonialism' to the current period of 'global coloniality (Gatsheni, 2013, p.3).

Coloniality continue to exist in the South African universities though now deeply entrenched in the organisational culture, norms and values. Also it is highly dependent on the individual academics and managers white and black alike, who are the bearers of such influence to turn the situation around. During the establishment of the RECs in some universities, people without any training and experience in dealing with research of ethical issues were given these roles and as a result were not taken seriously and they committed mistakes as they learned from scratch.

The continued existence of coloniality in the institutions of higher learning may well mean the subtle continuation of censorship of pivotal ideas by young black African scholars. It is against this assertion that the right to education can largely be enjoyed if accompanied by the academic freedom of students and staff (UN CESCR, 1999; Kori, 2016, p.46). Academic freedom also means right of people to freely express their opinions about system and institution within which they work, and this should enable fulfilment of their function without fear of repression and discrimination by the state or other interested parties (Kori, 2016, p.46).

Most African countries have their operations of RECs either remodelled or established on the bases of World Health Organisation (WHO) guidelines or in line with United States institutional review research oversight committees. Some countries have succumbed to the pressure exerted by research sponsors to establish human research oversight systems to reflect their institutional eligibility to receive research funding. Insistence of regulatory agencies of international research partners from their home countries had resulted in other countries introducing changes (Ndebele et al, 2014, p.4).

Black African student continue to be at the receiving end of limited opportunities and access to academic institutions and it is experienced at all academic levels and largely pronounced at postgraduate levels (Khoza-Shangase, 2019, p.45). This trend has a huge negative impact with the number of black students graduating at an undergraduate level, but mostly at postgraduate level. As a result, there is usually limited number of students who complete their postgraduate studies and consider academic path as a career compared to white and Indian races. It is likely that other races have systematically utilised institutions of higher learning to promote academic careers of their races. For instance, in a university where I read my master's degree, final year and honours students of Indian race were given preference to become research assistants to support departmental research endeavours compared to black African students. Very few black African students were given this opportunity and they had to be excelling in one way before being considered for those roles if their academic excellence could not be overlooked.

If a black African student made it to the academic space with outstanding publishing record, they suffered negation of their full professorship application after prolonged periods of consideration (Khoza-Shangase, 2019, p.44). The blocking of black African full professorship promotion is one of many challenges which include "*blocking access, progress and success of scholars within the academy*" (Khoza-Shangase, 2019, p.45).

This blockading of black African scholars is building from practices such ethical clearance that were once directly used to suppress dissenting voices of colonialization in South Africa's universities. The challenge continues to be posed by individuals entrusted with the responsibility to mentor and nurture young African intellectual talent where we observe academic jealous and deliberate sabotage, unfairly holding onto university resources to the exclusion of black Africans. There is lack of enthusiasm to provide leadership through quality supervision so that there is increase in number of black Africans to undertake and accomplish their research outputs.

INDIGENOUS AND EUROCENTRIC RESEARCH WAYS

The second Afrikaner phase of colonialism saw the establishment of higher institutions of learning that were based on their Afrikaner language, values and norms, aiming at matching and possibly surpassing the strength of their predecessors, the English. In all these phases, the victim was the African indigenous people where intellectual resources were left underdeveloped. Indigenous African people were not allowed to develop papers that were to explicitly expose apartheid regime hence some reports were censored.

This censorship of research was unethical and it continued in the post-apartheid South Africa especially when the academic exercise is done on behalf of government and certain information has potential of denting the image of the democratically elected government. This is in line with Fanon's black elite conception where in the absence of white colonialists, black elite continue to play the role of former white colonialist masters. Fanon asserted that *the colonized man is an envious man* (Fanon 1967, p.39) who looks with lust at white men's possessions to an extent of wanting to mimic him in every way. It is this lust that has made black elite to fail to bring about radical changes of the colonial system, instead they prefer reforms that will largely benefit them while fundamentals in academia suffer.

Black elite has followed in footsteps of their white British predecessors where certain research institutions seemed to be surfacing unpleasant facts about the social issues prevalent during certain epoch's rule. These institutions were closed down in favour of those establishments that are friendly to the government (Ratcliffe, 2001, p.111). They refused publication of research findings unless there are significant amendments in the research findings.

This unethical behaviour has contributed in continued social ills that are ravaging the poor especially the African indigenous people of South Africa because solutions cannot be found if the causes of such strife are distorted through censored research and academic findings. This practice is unethical and is inconsistent with the African way of practice which encourages fairness, humaneness and authenticity (Tuhiwai-Smith, 2012, p.xiii). Research from the African perspective should be action research oriented. The indigenous research agenda should be concerned with healing, reconciliation, developmental and capacity building (Tuhiwai -Smith,2012, p.xi). It is therefore promoting self-determination, hope, excitement and self-empowerment (Tuhiwai -Smith,2012, P.xiii).

The issues of respect including consideration of African ancestral issues, providing feedback to the community about the research that was conducted and the long term posture of the relationship with the communities which are the fundamentals of the indigenous research methods are not universally taken into consideration by the current ethical clearance processes in the South African universities.

INDIGENOUS UNIVERSITIES

It has been proven that first human being originated from the African continent (Pavid, 2018). Linked to this is evidence that the oldest universities in the world could be traced from the African continent including some of the basics of different scholastic disciplines have to some extent their origins from the African scholarship thoughts. For instance, the University of Qarawiyyin in Fes in Morocco was established in AD 859 while the Cairo's University of Al-Azhar in Egypt began its operations in AD 972. Furthermore, the Sankoré University in Timbuktu in Mali, was founded in the twelfth century (Ndlovu-Gatsheni).

While research is known to be closely linked to colonialism and European imperialism (Tuhiwai -Smith, 2012, p.1), the indigenous universities had their original research routines, processes and procedures that were later distorted by the colonialists. The chapter is asking questions such as how the ethical research consideration would have evolved in the African universities if coloniality flair were to be removed from the picture? How the academic African scholarship rigor would have developed and implemented for the promotion of African ethical practice based research? How the African research orientation would have been susceptible to various kinds of unorthodox African research methodologies because the African continent is arguably the bearer of all original thoughts as a mother of the world's first human beings?

African scholastic orientation and growth was redirected when artefacts and research material were stolen from the continent, only to being re-introduced in dominating ways, the so called westernisation of knowledge. The scholastic growth in Africa could have gone a long way to embrace local cultures and languages in higher institutions of learning. Hence the high moral ground even on the side of ethical clearance would have been greatly achieved.

For instance, Kaya, Kamwendo & Rushubirwa (2016, p.121) argued that "*development goals can only be achieved if knowledge and information is communicated effectively in the local language related to the cultural values of the people*". They further accused African indigenous languages marginalization in higher educational institutions seen as centres of knowledge production, as contributed to the African indigenous knowledge systems neglect in higher education. The indigenous art of orally transferring knowledge from one generation to the other using local languages has been undermined (Kaya, et. al, 2016, p.121). Oral transferring of knowledge has played a role in indigenous research, knowledge development and management as well as innovation for centuries. Indigenous languages combined with oral traditions of transferring knowledge are pivotal as anchors of African research and knowledge imparting for greater comprehension, learner enhancement of innovation and use of knowledge to solve community's immediate socio-economic challenges.

According to Alatas (2004) what exists in the modern epoch are institutions of higher learning modelled in Eurocentrism and are referred to as concept of 'the captive mind' to imply uncritical imitation of Western research paradigms within scientific intellectual activity (Kaya et al, 2016, p.122). The attempts by black African scholars to engrain their values and cultures into academia are met with invisible resistance. The entire neglect of African cultural context in academic ethical research processes is linked to the colonisation of the mind. When the mind is colonised there is stripping the colonised of their ancestral culture including indigenous languages and substituting it with Western culture (Kaya et al, 2016, p.122).

Transferring knowledge to students utilising colonial languages is tantamount to rejecting their own heritage while promoting the embrace as the human norm of the Western lifestyles, value systems and worldviews. Louis (2006) concurs that in African higher education, this rejection takes place in all the stages of the research process, that is, identification and selection of research problems, selection of methodology as well as methods of data collection. All these are guided by literature sources which are predominantly Western; the language used in the construction of research instruments and the dissemination of research findings is in most cases that of the colonisers. The research process becomes a struggle between the interests of the Western knowledge and Africa indigenous ways of knowing and value systems (Kaya et al, 2016, p.122).

AFRICAN MORAL STANDARDS

Indigenous people are keen to learn from research that addresses their needs (Zimmerman, 2005, p.307). As their consciousness is guided by questions of what value is the research, if it would not deal with indigenous people's socio-economic needs. When dealing with indigenous research requires understanding of culture's views of time or land, also their peculiar epistemologies about the past (Zimmerman, 2005, p.308). There are earlier theorists, who saw traditional institutions and knowledge as hindrances to development. However, this was a myth introduced by the colonisers so as to create black African intellectual self-doubt.

It should be noted that various cultures consist of different moral requirements (Mangena, 2012, p.1). While dialogue remains a two-way communication, it involves communication between vanhu (human beings) and spiritual beings such as midzimu (ancestral spirits) and musikavanhu (creator God) (Mangena, 2012, p.9). black Africans are attached to the departed and therefore Black African ancestral knowledge should be considered and recognised in the South African scholarly work.

Communal or group rationality means that kind of rationality where the opinions of the group take centre stage compared to the opinions of the individual (Mangena, 2012, p.10). Individual's importance is seen with regard to the way he or she participates in a group (Mangena, 2012, p.11). The African moral standards embrace communalism as opposed to individualism. For instance, if the current modern university ethical research process integrated black African way, it would promote research projects bearing academic credits that are undertaken by a group rather than individuals. Therefore, the entire Eurocentric orientation of Western academic morality and spirit is individualistic in nature and therefore un-African.

The uBuntu is what made black African people unique from many nations in that as part of bringing up children, they understood that they could not thrive in life without voluntary supporting others and allow voluntary support from others. Hence the proverb, 'umuntu ngumuntu ngabantu', which directly means 'a person is a person through other persons'. This is interpreted as whatever a person is, should be for the betterment of the community to which he or she belongs rather than seeing himself as an isolated being (Mangena, 2012, p.11). It is against this reason that the responsibility of raising children lies with the community rather than individual biological parents.

A person can adhere to the common moral position if his or her actions are in conformity with the community's agreed moral standards. In the case of sub-Saharan African cultures, the process of reaching a common moral position is dialogical (Mangena, 2012, p.12).

Lephalala (2012) considers uBuntu "*as a complex, elusive and multifaceted as it mirrors the multiple and shifting nature of the African society and human relationship*". In agreement with Lephalala (2012), Letseka (2008) regards uBuntu "*as a moral theory, an interactive ethic in which our humanity is shaped by our interaction with others*". uBuntu is antithetical to people growing to become hermits and discourages 'I', but encourages 'we'.

The African principle behind the Collective Finger Theory is that "*a thumb although is strong cannot kill aphids on its own, it would require the collective cooperation of the other fingers*" (Mbigi & Marre, 1995, p.110). Therefore, uBuntu values communalism, cooperation, respect, compassion, sharing of ideas and resources (Mbigi & Marre, 1995 p.110).

African society's moral orientation would be incomplete without the role of females who serve as first teachers of children. In the African society we refer to acquiring of mother tongue that has been suckled from the mother's breast. This perspective shows the early bonding between the mother and the baby that facilitates the initial teaching of the language to the infant by her mother. Furthermore, societal values, norms and culture is instilled at an infancy stage, which becomes important teaching that grooms children at an early stage to become uBuntu inclined individuals. Suda (1996) contends that:

female influence on the moral character of African society is arguably one of the most enduring marks in the history of humanity. Part of the reason for this is women's traditional care-giving roles, which put them in a unique and strategic position not only to produce and sustain life but also to help instil socio-religious values and morals in the family and society as a basis for establishing good and appropriate relationships between members (Suda, 1996, p.71).

Colonialisation and partition of Africa unleashed unprecedented psychological, cultural, political and economic changes in African societies (Ndlovu-Gatsheni, 2018, p.37). As a result, there is so much knowledge that had fallen through the cracks and need to be revived by acknowledging African-ness in research ethics and curriculum in higher education institutions. Moreover, black African students and academics should be allowed to explore this indigenous scholarly orientation so as to uncover intellectual richness of the African continent.

AFRICAN KNOWLEDGE AND REVEALING

Indigenous knowledge or African knowledge, here used interchangeably, is experiential knowledge based on a worldview and a culture that is basically relational. The spirit of the African worldview includes wholeness, community and harmony which are deeply embedded in cultural values. A person becomes human only in the midst of others and seeks both individual and collective harmony as the primary task in the process of becoming a true person (Sarpong, 2002). Like its people, acquisition of knowledge is collective and community oriented. Central to the African worldview is the strong orientation to collective values and harmony rooted in a collective sense of responsibility – a 'collective ethic' – which acknowledges that survival of the group derives from harmony through interdependence and interconnectedness (Mkabela, 2005; Sarpong, 2002).

A helpful example comes from a South African colleague who engaged rural indigenous women by means of focused group discussions in her quest to understand the concept of health from an indigenous perspective. She learned that for the rural indigenous women of Southern Africa health was basically about relationships. They believed that ill health occurred when relationships have broken down. In their indigenous thinking, even when restoration to health is sought through the biomedical approach, wholeness was possible only when broken relationships have been mended and restored because a person is healthiest when she or he is in harmony with others (Durie, 2004).

African knowledge, and its method of acquisition, has a practical, collective and social or interpersonal inclination. The works of Mpofu (2002) and Nsamenang (2006) show that indigenous conceptions of intelligence, for example, emphasise the practical, interpersonal and social domains of functioning and are quite differentiated from the cognitive 'academic' intelligence that dominates Western concepts of the construct. Likewise, learning for the African child is mostly peer oriented and participatory with less adult instruction (Pence & Nsamenang, 2008) which is consistent with a generally collective African self-concept (Mpofu, 1994).

As a knowledge system, it is characteristically oral and passed on from generation to generation in the context of community living and activities (Mkabela 2005).

Considering that culture is the 'lens' through which a person perceives, interprets and makes sense of his or her reality, if we speak of the inclusion of African indigenous knowledge in any investigation, it is referred to examination of African reality from the perspective of the African and not with the African on the periphery.

However, oral tradition, characteristic of indigenous knowledge systems is often, although erroneous, looked down upon relative to the written tradition. Partly because indigenous knowledge is mainly oral and not written, and partly because it is people-centered and sometimes not so easily 'measurable' (Emeagwali, 2003), It has therefore been mistaken by many as simplistic and not amenable to systematic scientific investigation. However, its rich complexities are found in the community ceremonies and rituals, namely, story-telling, proverbs, folktales, recitation, demonstration, sport, epic, poetry, reasoning, riddles, praise, songs, word games, puzzles, tongue-twisters, dance, music, and other education-centered activities (Ngara, 2007).

It is prudent to say that African indigenous knowledge is not without limitations. Some of these are the tenacious continuity of practices and beliefs that lack openness and flexibility to necessary or constructive changes. There is a tendency to forget that, just as any other type of knowledge, it is subject to change from economic, environmental and social forces (Tanyanyiwa & Chikwanha, 2011) and therefore its acceptance must be subject to critical observation and analysis. However, before the advent of Western methods of scientific inquiry African knowledge and methods have successfully guided its peoples in all spheres of life, this includes the realms of spiritual, social, educational, agricultural, political and economic (Pence & Nsamenang, 2008; Nsamenang, 2006; Mpofu, 2006). It is pivotal that African knowledge and methods of knowing must drive African research if it is to be meaningful to its peoples.

RESEARCH ETHICS AND ETHICAL CLEARANCE

Based on the secondary literature review of ethical clearance in South African universities, seem to be all perfect in the current juncture. Codes of practice in research provide guidelines that reinforce basic principles of both human rights and ethics; many aspects are legally enforceable (CCCU, 2014, p.2).

The unethical behaviour in handling ethical clearance might be resulting in excluding certain intellectual ideas that are not in sync with the neo-colonial establishment. For instance, there were disagreements by REC, negating community researchers' aspirations fuelled by motive to solve community challenge (Tuck & Guishad, 2013, p.10). The research needs propelled the community of researchers to aggressively collect data from small local shops who were illegally selling single cigarettes which had increased health risk for local youth and this suggestion was shot down by the research ethic committee on the pretext that university will not be involved in illegal activities of entrapping shop owners (Tuck & Guishad, 2013, p.10). This has become an example of how sometimes farfetched are the views of RECs in assessing and evaluating risk, respect, protection and benefit in research compared to the community researchers on the ground (Tuck & Guishad, 2013, p.10). The handling of ethical clearance may promote or retard intellectual progress amongst the black African indigenous people. The conceptual understanding of research within the African context has implications for the research ethics because in Africa all knowledge and ideas come from society and its inhabitants, the researcher becomes the coordinator and the documenter of such knowledge.

In pursuing black indigenous research ethics, new concepts should be developed while some concepts may be modified, utterly rejected and challenged leading to the creation of the new sets of concepts (Small, 2004, p.82). This should be done because many research concepts are derived from the worldview of eurocentrism and whiteness. With the trajectory pointing to decolonization, new black African and indigenous ethical research concepts should be developed.

The chapter argues that the subtle exclusion of indigenous black African intellectual work continued through the introduction of certain measures such as ethical clearance practices that seemed to exclude the advancement of indigenous African people's knowledge preservation, production and their intellectual growth. The worst situation is when some academics who lack uBuntu in their academic orientation and ethical scope to continue contravening the provisions of research ethical policies. The exclusionary practices are normalised in everyday life within academic institutions and such interactions in spaces that sustain exclusion are pivotal in creating hegemonic character of whiteness invisible (Kiguwa, 2019, p.18). These exclusions happen in full view of individuals in the academic institutions, yet in reality they "*effectively sustaining a closed network of who is allowed in and how decisions are made*" (Kiguwa, 2019, p.18).

CHALLENGES ASSOCIATED WITH RESEARCH ETHICS IN UNIVERSITIES

Ethical principles or Codes are a manifestation of expected behaviour amongst society and individuals. It associated to moral judgments aimed at guiding decision making and behaviour in peculiar situations. Therefore, ethical principles are connected to cultural values at a particular time in our history and are evolving as values and attitudes change. For instance, the prevailing dominant political ideology influences perception of societal and institutional moral compass, values and norms such as the period of colonialism where immorality was moderated and entrenched to be a norm. The immoral behaviour of colonial era inevitably permeated through to how ethical considerations were interpreted and applied in research at the time in history.

In conducting research, there is conflict between hastily conducting a study and ethical rules to follow in engaging such research in a respectful way to the subjects. On one hand, researchers are focused on developing knowledge which include the methodology utilised in research projects such as research protocols, subject selection, sample size, equipment, statistical analysis and personnel. On the other hand, researchers exercise ethos of respecting the research environment that demands attention to be exerted on appropriate utilisation of physical resources such as funds as well as animal and human subjects. In reality research considerations has gradually become a musk that hides immoral behaviour employed in research involving human beings. For instance, I argue that the observed unethical medical trials in the Durban area conducted by medical research institutions linked to higher institutions of learning have targeted vulnerable, poor, desperate, under-educated and unemployed young people who are largely unaware about their constitutional and human rights. The medical trails are conducted with a promise of meagre remuneration have resulted in jeopardising young people's future either through being physically infected or psychologically disturbed. Continuing attention to research conduct is needed in view of the fact that a meta-analysis of survey studies found that 30-74 percent of respondents' report that they have been involved in or observed inappropriate behaviour in the conduct of research (Fanelli, 2009).

Ethical considerations are an underlying instrument to determine a need for execution of research and methods for its implementation. The scientific research process affects those who are subjects as well as perception of public regarding scientists and science. Further studies may be influenced by research outcomes and in turn influence legislation and public policy. If there is doubt in scientific integrity, public belief in findings and confidence may be eroded. It is against these reasons that ethical principles should be followed when conducting research while putting at the forefront honesty, transparency and competency as one deals with research subjects.

There is a claim that many indigenous communities have difficulty in handling processes associated with researcher's request for permission given its unfamiliarity, complexity and expression in legal like terms (Zimmerman, 2005, p.304). Sometimes they associate such signing of documents with acts of fraud. There was also interrogation of western research paradigms, urging culturally appropriate practices and the conscious development of indigenous peoples as researchers (Linda Tuhiwai Smith; 1999; Thaman, 2003, p.4). Knowledge inherited from the academy is often regarded as inevitable, natural, and usually accepted as 'absolute truth', integrate such knowledge into culture of organizing and managing work, lives and ourselves.

Researchers and academics produce 'knowledge' through the 'truths' that are delivered and presented (Thaman, 2003, p.5).

CONCLUSION

Learning recognises the energy and potential within each person and tries to empower them to make their full contribution (Schenck, Nel & Louw, 2015, p.99). Hence the chapter interrogated the current ethical clearance processes in the South Africa's academic institutions and possibilities of being utilised to slow down African black intellectual and academic progress. In historically white universities research ethics committees should be widely representative of black African academics and women to avoid unscrupulous activities. The 'unethical clearance of the soul' phenomenon should be minimised by closely monitoring academic supervisors and demerit system should be introduced as part of the performance management system. The total number of students enrolled for supervision per lecturer should be compared to the number of students who finished their research on time. The bureaucratic research ethical clearance process should be simplified further. uBuntu should be developed further not only as underpinnings of South Africa's philosophy, but should be deeply entrenched in strategy and operations of all government institutions including institutions of higher learning and their RECs. Further investigation should be made not only of RECs, but of medical research bodies who may be continuing to practice unacceptable research ethical practices with a view to make concrete recommendations of incorporating Afrocentric values and culture in research ethical considerations.

Institutional cultures and learning environments should be transformed in order to be easier the process of incorporating Africanised ethical research considerations. For instance, there should be contemplation to credit research and its processes that infuses innovative African culturally based methodologies. Furthermore, educational credits should be given to group research as it embraces communalism and Afrocentricity as opposed to Eurocentric oriented individualism. Long term relationship with the community within which research is conducted should be encouraged.

RECs should put emphasis on usefulness of research such that disadvantaged communities stand to immediately benefit from it. Such utility and expediency of research should also contribute towards addressing South African development challenges. Localisation and usefulness of research should also be central in awarding scholarships funded from the South African fiscus rather than research investment made for the benefit of foreign countries at the expense of local financial resources. There should be revival of entrepreneurship and lifelong learning oriented education rather than education whose only focus is to create a workforce to support labour market.

It is clear that the African scholarly knowledge has continuously been undermined. Not only the Eurocentric knowledge currently dominates the higher institutions of learning, but also cultural values and norms. These ingredients were deeply entrenched during the colonial and apartheid era in South Africa and they also influenced processes and procedures that were put in place to facilitate research ethical clearance. I argue that in cases where apartheid higher institutions of learning were unable to out-rightly prevent black Africans to access higher education, they however succeeded to subtle sabotage their academic endeavours in a number of ways, including utilisation of research ethical clearance. This happened because few of the white racist and patriarchal men who dominated academia during the colonial epoch wanted to be associated with staunch support of discrimination especially in academia. They resorted into clandestine practices because some of their white counterparts truly believed in dismantling the apartheid machinery that was connected to infringement of academic freedom and perpetuation of censorship policies in academia and countrywide. Some white academic liberals became state enemies because they stood firm against these injustices. Nevertheless, black African women and men suffered in colonial universities both at a level of being students and academics, hence there was limited academic publication by African black intellectuals compared to whites partly due to delaying and blockading tactics employed during research ethical clearance processes. This trend continued to some extent during the post-apartheid era hence "*fees must fall*" and "*Rhodes must fall*" movements constituted brave young men and women who were able to expose the continued injustices in academia while the black elite perpetuated the myth that democratic constitution of South Africa is continuation of the freedom charter. Meaning that access to education and academic freedom is embedded in South African constitution. The students of these aforementioned movements boldly questioned continued side-lining of black African youth from accessing expensive education because of their working class background and poverty. Also these movements challenged the continued existence of untransformed academic landscape on curriculum and administrative grounds. A clear victory was witnessed with the pronouncement of free education policy by the former state president Mr Jacob Zuma. However, the victory is uncertain with regards to the transformation of especially historically white universities. Black African intellectuals and students continue to be victims of invisible academic missiles aimed at breaking their intellectual integrity and stamina. The incorrect use of research ethical clearance processes and procedures to further the aspirations of former racist and patriarchal academic bureaucracy in colonial universities will not be discounted, especially as it is likely to have become a contemporary and subtly scramble for the elusive fulfilment of coloniality in our South African shores.

ACKNOWLEDGMENT

This project received no specific grant from any funding agency in the public, commercial or not-for profit sectors.

REFERENCES

Akoojee, S., & Nkomo, M. (2007). Access and quality in South African higher education: the twin challenges of transformation. *SAJHE, 21*(3), 385-399.

Alatas, S. H. (2004). The captive mind and creative Development. In P. N. Mukherji & C. Sengupta (Eds.), *Indi-geneity and Universality in Social Sciences: A South African Response* (pp. 24–31). Sage.

Beale, M. A. (1998). *Apartheid and University Education, 1948-1970.* Published Phd thesis: Obtained from. http://wiredspace.wits.ac.za/jspui/bitstream/10539/20610/1/Beale%20Mary%20Alice._Apartheid%20and%20university%20educati_001.pdf

Beckmann, J. (2016). Exclusion and Education in South Africa: An Education Law Perspective of Emerging Alternative Understandings of Exclusion. *Education Provision to Every One: Comparing Perspectives from Around the World BCES Conference Books, 14*(1), 295-302

Bevan, S. (2006). *White farmer stands trial for shooting a black boy he 'mistook for a dog'*. Obtained from https://www.telegraph.co.uk/

Bonilla-Silva. (2006). *Racism without racists: Color-blind racism and the persistence of racial inequality in the United States* (2nd ed.). Rowman & Littlefield Publishers, Inc.

Bourdieu, P., & Passeron, J.C. (1990). Theory, culture & society. Reproduction in education, society and culture (2nd ed.; R. Nice, Trans.). Sage Publications, Inc.

Cajetas-Saranza, R. (2015). Approaches to Integrating Multicultural Perspectives in PNU-Mindanao: Basis for Multicultural Education Program. *International Peer Reviewed Journal, 16*, 37–55. doi:10.7718/iamure.ije.v16i1.1122

Canterbury Christ Church University (CCCU). (2014). *An introduction to ethics issues and principles in research involving human and animal participants*. CCCU.

CHE (Council for Higher Education). (2017). Decolonising the curriculum: Stimulating debate. *Briefly Speaking, 3*, 1-12.

Cloete, F., & Auriacombe, C. (2019). Revisiting decoloniality for more effective research and evaluation. *African Evaluation Journal, 7*(1). doi:10.4102/aej.v7i1.363

Connell, R. W. (1993). Schools and Social Justice. Temple University Press.

Dastile, N.P., & Ndlovu-Gatsheni, S.J. (2013). Power, Knowledge and Being: Decolonial Combative Discourse as a Survival Kit for Pan-Africanists in the 21St Century, *Alternation, 20*(1), 105 - 134.

De Wet, K. (2010). 2010, The Importance of Ethical Appraisal in Social Science Research: Reviewing a Faculty of Humanities' Research Ethics Committee. *Journal of Academic Ethics*, *8*(4), 301–314. doi:10.100710805-010-9118-8

Delmas, A. (2011). From travelling to history: An outline of the VOC writing system during the 17th century. In A. Delmas & N. Penn (Eds.), *Written Culture in a Colonial Context: Africa and the Americas 1500–1900*. UCT Press. doi:10.26530/OAPEN_628137

Dempsey, P., & Dempsey, A. (1992). *Trials of War Criminals before the Nuremburg Military Tribunals: The Medical Case 1948*. Jones and Bartlett.

Do Vale, H. F. (2016). Explaining Education Reforms and Decentralization in Brazil, South Africa, South Korea, and Spain. *Lex localis - Journal of Local Self-Government*, *4*(3), 591–612. doi:10.4335/14.3.591-612(2016)

Durie, M. (2004). An Indigenous model of health promotion. *Health Promotion Journal of Australia, 15*(3), 181–185. doi:10.1071/HE04181

Emeagwali, G. (2003). African indigenous knowledge systems (AIK): Implications for the curriculum. In T. Falola (Ed.), *Ghana in Africa and the world: Essays in honour of AduBoahen*. Africa World Press.

Fanelli, D. (2009). *How Many Scientists Fabricate and Falsify Research? A Systematic Review and Meta-Analysis of Survey Data*. doi:10.1371/journal.pone.0005738

Fanon, F. (1967). The wretched of the earth. New York: Grove Press.

Fanon, F. (1968). *The Wretched of the Earth*. Grove Press.

Freire, P. (1972). *Pedagogue of the oppressed*. Penguin.

Garriott, P. O., Reiter, S., & Brownfield, J. (2016). Testing the Efficacy of Brief Multicultural Education Interventions in White College Students. *Journal of Diversity in Higher Education, 9*(2), 158–169. doi:10.1037/a0039547

Gatsheni, N. (2013). *Coloniality of Power in Postcolonial Africa Myths of Decolonization*. Academic Press.

Gelling, L., & Munn-Giddings, C. (2011). Ethical review of action research: The challenges for researchers and research ethics committees. *Research Ethics Review, 7*(3), 100–106. doi:10.1177/174701611100700305

Grosfoguel, R. (2007). The Epistemic Decolonial Turn. *Cultural Studies, 21*(2-3), 211–223. doi:10.1080/09502380601162514

Heleta, S. (2016). Coloniality persist in our universities and we must urgently decolonise. *Mail & Gaidian*. Obtained from https://mg.co.za/article/2016-11-18-00-coloniality-persists-in-our-universities-and-we-must-urgently-decolonise

Higher Education South African. (2014). *South African Higher Education in the 20th Year of Democracy: Context, Achievements and Key Challenges HESA presentation to the Portfolio Committee on Higher Education and Training Cape Town*. Retrieved October 17, 2016, from https://pmg.org.za/committeemeeting/ 17037/

Kallaway, P. (2002). *The history of education under apartheid, 1948–1994: the doors of learning and culture shall be opened*. Pearson.

Karabel, J. (1996). Towards a Theory of Intellectuals and Politics. *Theory and Society, 25*(2), 205–233. doi:10.1007/BF00161141

Kaya, H. O., Kamwendo, G. H., & Rushubirwa, L. (2016). African Indigenous Languages in Higher Education. *Studies of Tribes and Tribals, 14*(2), 121–128. do i:10.1080/0972639X.2016.11886739

Khoza-Shangase, K. (2019). Intellectual and Emotional Toxicity: Where a Cure Does Not Appear to Be Imminent. In G. Khunou, H. Canham, K. Khoza-Shangase, & E. D. Phaswana (Eds.), Black Academic Voices: The South Africa Experiences. Academic Press.

Khunou, G. (2019). Black in the academy: Reframing knowledge, the knower, and know. In G. Khunou, H. Canham, K. Khoza-Shangase, & E. D. Phaswana (Eds.), Black Academics Voices: The South African Experience. Academic Press.

Kiguwa, P. (2019). Negotiating the academy: Black bodies 'out of place. In G. Khunou, H. Canham, K. Khoza-Shangase, & E. D. Phaswana (Eds.), Black Academic Voices: The South Africa Experiences. Academic Press.

Kim, S. Y. H., Ubel, P., & De Vries, R. (2009). *Pruning the regulatory tree*. PubMed., doi:10.1038/457534a

Kori, E. (2016). Challenges to academic freedom and institutional autonomy in South African universities. *International Journal of Teaching and Education, IV*(1), 45–53. doi:10.20472/TE.2016.4.1.004

Le Roux, E.H. (2013). *Between Complicity and Resistance: A Social History of the University Presses in Apartheid South Africa*. Published Phd thesis: Obtained from https://repository.up.ac.za/bitstream/handle/2263/25429/00front.pdf

Lephalala, M. M. K. (2012). *Between Past and Present: Conflict in schools, an Ubuntu Perspective*. University of South Africa Press.

Letseka, M. (2012). In Defense of *Ubuntu*, Studies in Philosophy and Education. *International Journal (Toronto, Ont.), 31*(1), 47–60.

Louis, R. P. (2006). Can you hear us now? Voices from the margin: Using indigenous methodologies in geographic research. *Geographical Research, 45*(2), 130–139. doi:10.1111/j.1745-5871.2007.00443.x

Mangena, F. (2012). Towards a hunhu/ubuntu dialogical moral theory. *Phronimon, 13*(2), 2012.

Mbembe, A. J. (2016). Decolonizing the university: New directions. *Arts and Humanities in Higher Education, 15*(1), 29–45. doi:10.1177/1474022215618513

Mbigi, L. (1995). *uBuntu: A Rainbow Celebration of Cultural Diversity*. uBuntu School of Philosophy,Tshwane.

Mbigi, L., & Maree, J. (1995). *uBuntu: The Spirit of African Transformation Management*. Sigma Press.

Mkabela, Q. (2005). Using the Afrocentric method in researching indigenous African culture. *Qualitative Report, 10*(1), 178–189.

Mlatsheni, C. (2012). The challenges unemployment imposes on youth. In H. Perold, N. Cloete, & J.Papier (Eds.), Shaping the future of South Africa's youth: Rethinkingpost-school education and skills training (pp. 31-41). Cape Town: Centre for Higher Education Transformation (CHET).

Mpofu, E. (1994). Exploring the self-concept in an African culture. *The Journal of Genetic Psychology, 155*(3), 341–354. doi:10.1080/00221325.1994.9914784 PMID:7964660

Mpofu, E. (2002). Indigenization of the psychology of human intelligence in SubSaharan Africa. In W.J. Lonner, D.L. Dinnel, S.A. Hayes, & D.N. Sattler (Eds.), *On-line Readings in Psychology and Culture*. Center for Cross-Cultural Research, Western Washington University.

Msila, V. (2015). Ubuntu, Shaping the current workplace with (African) wisdom. Knowres Publishing (Pty) Ltd.

Mthembu, N. C. (2017). *Black African Township youth survival strategies in post-apartheid South Africa: A case study of the KwaMashu Township within eThekwini Municipality* (PhD Thesis). University of South Africa, Tshwane.

Ndebele, P., Mwaluko, G., Kruger, M., Ouwe, O., Oukem-Boyer, M., & Zimba, M. (2014). History of Research Ethics Review in Africa. In Research Ethics in Africa: A Resource for Research Ethics committees. Sun MeDIA Stellenbosch.

Ndlovu-Gatsheni, S.J. (2017). *The emergence and trajectories of struggles for an 'African university': The case of unfinished business of African epistemic decolonization*. Academic Press.

Ndlovu-Gatsheni, S. J. (2018). Genealogies of Coloniality and Implications for Africa's Development. In V. Gumede (Ed.), *Inclusive development in Africa: Transformation of Global Relations*. doi:10.2307/j.ctvh8r26b.9

Ndlovu-Gatsheni S.J., (n.d.). Global Coloniality and the challenges of crating African futures. *Strategic Review for Southern Africa, 36*(2).

Ngara, C. (2007). African ways of knowing and pedagogy revisited. *Journal of Contemporary Issues in Education, 2*(2), 7–20.

Nsamenang, A. B. (2006). Human ontogenesis: An indigenous African view on development and intelligence. *International Journal of Psychology, 41*(4), 293–297. doi:10.1080/00207590544000077

Oliphant, A. W. (2000). From Colonialism to Democracy: Writers and Publishing in South Africa. In N. Evans & M. Seeber (Eds.), *The Politics of Publishing in South Africa. Scottsville*. University of Natal Press.

Pavid, K. (2018). *Rethinking our human origins in Africa, Natural History Museum*. Obtained from: https://www.nhm.ac.uk/discover/news/2018/july/the-way-we-think-about-the-first-modern-humans-in-africa.html

Peltier, C. (2018). An application of Two-Eyed Seeing: Indigenous Research Methods With Participatory Action Research. *International Journal of Qualitative Methods, 17*(1), 1–12. doi:10.1177/1609406918812346

Pence, A., & Nsamenang, B. (2008). *A case for early childhood development in subSaharan Africa*. Working Paper No. 51, Bernard van Leer Foundation, The Hague, Netherlands.

Pinar, W. F. (2009). *Understanding Curriculum: An Introduction to the Study of Historical and contemporary curriculum discourses* (W. F. Pinar, W. M. Reynolds, P. Slattery, & P. M. Taubman, Eds.). Peter Lang.

Pityana, B. (2008). Reflections on 30 years since the death of Steve Biko: A legacy revisited. In C. W. du Toit (Ed.), *The legacy of Stephen Bantu Biko* (pp. 1–14). UNISA.

Pityana, N. (1981). The Black Consciousness Movement and Social Research. In J. Rex (Ed.), *Apartheid and Social Research* (pp. 161–184). UNESCO.

Ratcliffe, P. (2001). Race' and Ethnicity Research in Britain: Key Ethical and Political Considerations. In P. Ratcliffe (Ed.), *The politics of social science research: race, ethnicity, and social change*. Palgrave. doi:10.1057/9780230504950_6

Resnik, D. B. (2015). *What is Ethics in Research & Why is it Important?* National Institute of Environmental Health Sciences: https://www.niehs.nih.gov/research/resources/bioethics/whatis/index.cfm

Rodney, W. (1972). *How Europe underdeveloped Africa*. Bogle-L'Overture Publications.

Sarpong, P. (1991). *Ghana in retrospect: Some aspects of the Ghanaian culture*. African Books Collective Ltd.

Sarpong, P. (2002). *Peoples differ: An approach to inculturation in evangelization*. SubSaharan Publishers.

Schenck, R., Nel, H., & Louw, H. (2015). *Introduction to participatory community practice*. Unisa Press.

Small, S. (2004). Researching 'mixed-race' experience under slavery: concepts, methods and data. In Researching Race and Racism. Routledge.

Strydom, F., Kuh, G., & Mentz, M. (2009). Enhancing success in South Africa's higher education: measuring student engagement. *Acta Academica, 42*(1).

Suda, C. (1996). *The Centrality of Women in the Moral Teachings in African Society. Nordic Journal of African Studies*, *5*(2), 71–82.

Sue, D. W., & Sue, D. (1990). *Counseling the Culturally Different: Theory and Practice* (2nd ed.). Wiley.

Tanyanyiwa, V. I., & Chikwanha, M. (2011). The role of indigenous knowledge systems in the management of forest resources in Mugabe area, Masvingo, Zimbabwe. *Journal of Sustainable Development in Africa, 3*(3), 132–149.

Thaman, K. H. (2003). Decolonizing Pacific Studies: Indigenous Perspectives, Knowledge, and Wisdom in Higher Education. *The Contemporary Pacific, 15*(1), 1–17. doi:10.1353/cp.2003.0032

Thompson, J.B. (1984). *Studies in the Theory of Ideology, Social and political theory*. Polity Press.

Trust, S. (2019). Knowledge Valorisation for Inclusive Innovation and Integrated African Development. In S. Oloruntoba & M. Muchie (Eds.), *Innovation, Regional Integration, and Development in Africa. Advances in African Economic, Social and Political Development*. Springer. doi:10.1007/978-3-319-92180-8_7

Tuck, E., & Guishard, M. (2013). Un-collapsing ethics: Racialized sciencism, settler coloniality, and an ethical framework of decolonial participatory action research. In T. M. Kress, C. Malott, & B. Porfino (Eds.), *Challenging status quo retrenchment: New directions in critical research* (pp. 3–28). Information Age Publishing.

Tuhiwai-Smith, L. (2012). Decolonizing methodologies, Research and Indigenous People's (2nd ed.). Ottago University Press.

Unite Nations. (1999). Statement of the UN Committee on Economic, Social and Cultural Rights to the Third Ministerial Conference of the World Trade Organisation (Seattle, 30 November to 3 December 1999).

University of Alabama. (2007). *U.S. Department of Health & Human Services*. https://obssr.od.nih.gov/wp-content/uploads/2016/05/Ethical-Challenges.pdf

Wa Thiong'o, N. (1981). *Decolonizing the mind: The politics of language in African literature*. East African Educational Publishers.

Welsing, F. C. (1991). *The ISIS Papers: The keys to the colours*. Third World Press.

WHO. (2009). *Research ethics committee: basic concepts for capacity building*. WHO Press.

Zimmerman, L. J. (2005). First be humble: working with indigenous people and other descendant communicates. In Indigenous Archaeologies: Decolonising Theory and Practice. Academic Press.

KEY TERMS AND DEFINITIONS

Coloniality: Is a long-standing patterns of power that occurred ensuing from colonialism. These designs further define knowledge production, culture, labour, and inter-subjective relations.

Ethical Clearance: Is the process set by higher institute of learning and presided over by the research ethics committee to holistically regulate research projects to ensure their design and execution plan are ethically sound and will render anticipated results.

Ethics: Are considered to be framework for comprehending right or wrong while governing person's behaviour during conducting of research activity.

Research Ethics Committee: A group of individuals who undertake the ethical review of research protocols involving humans while applying agreed upon ethical principles.

Chapter 3
A Decolonial Curriculum Is Everything: An Afrocentric Approach

Zingisa Nkosinkulu
University of South Africa, South Africa

ABSTRACT

This chapter seeks to map how indigenous people and their indigenous knowledge systems are the most researched and written about in the world, yet they are the least understood. The curriculum of the empire and its scientific explanation justified how indigenous knowledge systems should be approached and viewed as well as who has the authority to justify; hence, indigenous knowledge systems were justified as inferior and not worthy of the standard of European knowledge system. In this chapter, Frantz Fanon's thought will be deployed to illustrate how this division of knowledge justifies the perpetuating dehumanisation of indigenous people under the mask of modernisation and globalisation. By deploying decoloniality, Afrocentricity, and Fanonian thought, this chapter seeks to challenge this curriculum that is based on the history of the conquest of Africa that positioned Africa only as a cradle of slaves and the black bodied as created by God only for the benefit of the Europeans.

INTRODUCTION

Since I was born in the Antilles, my observations and my conclusions are valid only for the Antilles—at least concerning the black man at home. Another book could be dedicated to explaining the differences that separate the Negro of the Antilles from the Negro of Africa. Perhaps one day I shall write it. Perhaps too it will no

DOI: 10.4018/978-1-7998-1249-4.ch003

longer be necessary—a fact for which we could only congratulate ourselves (Fanon [1952] 2008, p.1).

Frantz Fanon's perspective is instrumental to the understanding of the curriculum as a tool that divides the coloniser from the colonised in the colonial/Eurocentric/anti-black world. Fanon's lived experience is testament to this division that influenced his life and thoughts about the existence of the indigenous people that he refers to as the colonised or natives in his writings. Indigenous people and their indigenous knowledge systems are the most researched and written about in the world, yet they are still the most misrepresented. This is why Afrocentricity is a suitable paradigm to decolonize the curriculum in relation Fanon's thought. Afrocentricity is a paradigm that positions the Africana subject as the center of hers or his world, it allows the agency and ability to think form an African-centered perspective (Asante 2007; Owusu-Ansah & Mji, 2013). In this regard, Fanon's thought is relevant to the understanding of this epistemic division as something that materialized deliberately for the structure of the "empire" (Hardt & Negri 2000) that was intended to restructure the world to the image of the European logic. The knowledge systems and way of living for colonised people was the fundamental base of creating what transpired to be the curriculum of the empire, with research approaches that are being deployed in academia today. Most research approaches that are being used in the curriculum are based on research and methodologies that were developed and used towards the contribution of positioning black people as objects intended to be owned, used and studied. Under these research approaches, the colonised people's existence and their way of life was studied and used to create disciplines such as anthropology, ethnography, sociology and research methodologies that were used to justify indigenous people as inferior. While other academic disciplines such as science, art and archaeology as well as biology and history were used as a weapon to justify dehumanisation of indigenous people, and the classification of their knowledge systems as primitive and barbaric.

This epistemic division created and justified the existence of what is commonly known as the Global South in which Africa as the victim of colonisation is located under, as the primitive side of knowledge. Fanon understood the language of the centre and periphery that demonstrated these divisions that made the periphery a no man's land, as the "zone of nonbeing" (Fanon [1952] 2008, p.2). The zone of none being is a place without knowledge, a place without history, without spirituality and most importantly a place without humans. As history tells us about what resulted to Africa being named a dark continent, a name that embodies a perception of many fathers of European knowledge such as Hegel. This perception is based on the history of the conquest of Africa that positioned Africa only as a cradle of slaves and black bodies as created by God only for the benefit of European pleasures and development. Narratives of history and of the conquest depicts to us how colonial research and methodologies were supported by the European fathers of knowledge. Fanon takes a different position from the European fathers of knowledge to say that the colonised life is life in the zone of nonbeing, there are beings and there is a different humanity. This chapter takes Fanon's position of thinking beyond the curriculum of the empire by adopting an Afrocentric approach.

LITERATURE REVIEW

Education and its curriculum in most colonised countries is Western, or put differently, after colonisation quality education was only for white people specifically during the apartheid era in South Africa. However, the politics of the apartheid system in South Africa did not exist in a vacuum, their genealogy is connected to all the struggles against colonisation of indigenous societies who were colonised or westernised. Hence, Aimé Césaire posed a very important question that proves to be relevant to the understanding of how the curriculum as the source of knowledge, was appropriated by the colonial master as a tool that builds the empire. However, before this question is posed and answered in this chapter, it will be of great assistance to understand two things, the empire and the curriculum of the empire. As explained by Hardt and Negri (2000, p. vii), "Empire is the political subject that effectively regulates these global exchanges, the sovereign power that governs the world". The empire is the reason why everything is the way it is, it is the reason why the curriculum was and still is appropriated, it is the reason why colonisation, slavery, capitalism, modernisation and globalisation were created as systematic forces that rules the world. "The passage to Empire emerges from the twilight of modern sovereignty" (Hardt & Negri 2000, p. xiii) to position Europe in the centre of the world by reconfiguring ways of building the Eurocentric world, an anti-black world. The hidden thread that is imbedded in this anti-black world highlights that anti-blackness is the fuel that keeps running the machineries that develops what I will term here the curriculum of the empire. The curriculum of the empire is a Eurocentric-asymmetric curriculum that approaches teaching and learning from a westernised perspective. It is a curriculum of lies and distortion, it bounds the Africana subject while it liberates the European subject. It is a curriculum that does not speak to the experience of the Africana subject, it closes indigenous knowledge system outside the curriculum, while it transforms the Africana subject into a black subject as what appears as hyper-slaves. An educated slave of the new system whose skin color precedes being and competence. The curriculum of the empire as one of the vital instruments that is utilized in order to spread and teach the knowledge of the empire is something that most colonised people have in common. It is a curriculum that renders them incompetence.

The curriculum of the empire can be explicated as "Western knowledge has been spread globally and, therefore, it is in all of us who went through at least secondary education" (Mignolo in Gaztambide-Fernández 2014, p.201). However, it has now become an ontological software that is shaping our everyday life because, the "empire is materializing before our very eyes" (Hard and Negri 2000, p. vii). It is in the materialisation of the empire that the world was divided into two oppositional categories namely; world and empire, centre-periphery, master and slave, rich and poor, colonised and coloniser based on black and white as skin color, as well as whiteness and blackness. It is through this Manichean structure of the world that the empire is born as Fanon articulates, "there is a zone of nonbeing, an extraordinarily sterile and arid region, an utterly naked declivity where an authentic upheaval can be born" (Fanon [1952] 2008, p. 2). Following Fanon's remark, it is assumed that there is also a zone of being that Fanon does not mention, it is a space of whiteness, but the empire is born at the cost of the zone of nonbeing, a place of blackness. It is from this place that the colonised people reside, a place where the empire mines knowledge to create its curriculum for the disciplines of knowledge. According to Smith (1999, p. 58-59), p.

Knowledge was also there to be discovered, extracted, appropriated and distributed. Processes for enabling these things to occur became organized and systematic. They not only informed the field of study referred to by Said as 'Orientalism' but other disciplines of knowledge and 'regimes of truth'. It is through these disciplines that the indigenous world has been represented to the West and it is through these disciplines that indigenous peoples often research for the fragments of ourselves which were taken, catalogued, studied and stored.

Judging by Smith's remark, the curriculum of the empire is a curriculum of appropriation, exclusion, single perspective of the world, dehumanisation, epistemic parasite, colonisation and westernisation of the world, it is a curriculum of fragmentation that feeds from what Gordon (2014) calls "disciplinary decadence". The curriculum of the empire was created from traveling experiences, observations, orientation, naming and recording of life, based on dairies and data collection of the colonial master. The creation of the curriculum of the empire in this manner resulted to it being the stone with which the epistemic walls between disciplines were constructed. These epistemic walls produced "an inward path of disciplinary solitude eventually leads to … disciplinary decadence. This is the phenomenon of turning away from living thought, which engages reality and recognises its own limitations, to a deontologised or absolute conception of disciplinary life" (Gordon 2006, p.86). Disciplinary decadence is based on the "worldview" (Oyewumi 1997) that operates only by the sense of sight, it is a perspective that informs how people view the world based from the Eurocentric logic, it is a view that is grounded on coloniality, it is what I can call coloniality of vision. The concept of coloniality of vision is grounded on the notion of "bad faith (mauvaise foi)" (Sartre [1957] 2003, p. 71, Gordon 1995) and anti-blackness. It does not see the Africana subject as a competent contributor of knowledge and content that can provide the world with a "new humanism" (Fanon 0000). Bad faith means that the European conquer viewed the African and concluded its black, therefore it is inferior, it is black therefore it is animal like, it is lacking a soul, it is not trust worthy, it is their property and most importantly, lets feed of them for our curriculum.

The European presence as the only one prominent in the curriculum of the empire means any other form of knowledge is questioned if whether worthy or not, to be documented as part of the curriculum. This is thinking from the position of the coloniality of vision and the curriculum of the empire that "the discipline becomes, in solipsistic fashion, the world. In that world, the main concern is the proper administering of its rules, regulations, or, as Fanon argued, (self-devouring) methods" (Gordon 2006, p.424). This world that the disciplines projected is a world with multiple faces than what it simply is, it is an anti-black world, a Manichean world, a world that is based on imperialism, it is a world that became an empire. It is an empire of which is created by the creation of colonies. The curriculum of the empire was filtered to the rest of the colonies by missionaries to influence how black people must relate to knowledge, being and power. Through the lense of the Fanonian thought, the curriculum of the empire meant there are "two poles of a world, two poles in perpetual conflict, p. a genuinely Manichean concept of the world" (Fanon [1952] 2008, p.31). This epistemic division was intentionally created to be the operating system of the empire, "a world divided into compartments, a motionless, Manicheistic world, a world of statues, p. the statue of the general who carried out the conquest, the statue of the engineer who built the bridge; a World which is sure of itself, which crushes with its stones the backs flayed by whips, p. this is the colonial world" (Fanon [1952] 2008, p. 50-51). The empire is a colonial world that is divided by colonies where the curriculum of the empire has become an epistemic statue of European presence. The European presence was something that manifested in many different and hidden ways, its footprint was stamped on top of every good discovery in the world. Most importantly, the European presence was imbedded in the mind of the colonised through the curriculum of the empire as the sign of modernity.

The Case of South African Experience

In the case of South Africa as a colony, it is amongst the list of colonial heirs whom the colonial grandfathers left their presence as legacy because it meant that it should exist forever. This settler presence extends as far as the social and political life of the South African people, it can be found on the architecture of buildings, it is on the names of streets and places, it is eaten as food diet and it is seen inside the dreams of the colonised people's mind about their free existence. In this regard Yountae (2017, p.97) indicates that, "the existential impasse of the black person lies, for Fanon, in the fact that his or her existence unfolds in the "zone of nonbeing," a state of perpetual curse". Fanon is a product of the curriculum of the empire who refused to be tamed and rather used its tools to find his way out the colonial subjection. The danger of the curriculum of the empire is that it "brings about essential mutations in the consciousness of the colonized, in the manner in which he perceives the colonizer, in his human status in the world" (Fanon 1965, p. 69). The colonised become hypnotised by the romantic presence of the coloniser in the curriculum who is always depicted as the victor of wars, founder of places and the god of knowledge. The effects of this hypnotism are that it makes the colonised have a different perception of the self and the coloniser. This curriculum hypnotism is similar to the frog that is experiencing paralyses under the snake's watch. In South Africa, the colonised get hypnotised by the presence of the coloniser in the curriculum of the empire and get eaten by modernity by being closed out of the ontological infrastructure that is supposed to grant access to the rights to live.

In the case of South Africa, the issue of the presence of the coloniser in the curriculum of the empire as the victor and the only father of knowledge, is mostly felt by the colonised and its contestation is manifested into resistance movements such as Fees Must Fall, Rhodes Must Fall and Decolonising the Westernised University. South Africa shares a lot in common with many ex-colonised places, as a result, its struggles are linked to similarities of most of the strategies that are taken to combat against imperialism. The movements Fees Must Fall, Rhodes Must Fall and Decolonising the Western University qualify as responding to imperialism, they do present these struggles. However, what these movements have in common is what is fundamentally important, it is what Fanon calls the colonised to do, it is being created by modernity and resisting against modernity. According to Ngugi wa Thiong'o (1993, p.22), "the modern world is a product of both European imperialism and of the resistance waged against it by the African, Asian, and South American peoples". The significance of these movements lies behind embodying the position of the smoke in the English idiom simply paraphrased as "where there is smoke, there is fire". If that's the case, the question that follows is where or what is the fire in the case of these South African student movements of decolonisation? What can be discovered if the fire is explored?

The curriculum of the empire is part of the fire, it is the reason why the aforementioned movements of decolonisation erupted in South Africa. These movements show that there is a problem that needs to be addressed. the building of the empire created these problems as problems that have been long in the making since the collapse of the idea of the world. It can be argued here, the idea of the world "is dead", by world here I don't mean the world in a physical and geographical sense, rather as a conceptual and epistemic place, an open place, a place without borders, where everybody has a place, an open-ended place. When the word "world" is uttered what comes into mind is a globe, it is planet earth, a free and open space, a place for every form of being. The world remained a world until capitalism introduced imperialism to transform it into an empire, during this transformation "the power of the colonialist message, the systems used to impose it and present it as the truth were such that most of the time the colonized had only his own increasingly overshadowed inner conviction to oppose" (Fanon 1965, p.76). The curriculum of the empire has subliminal messages that transforms the colonised subject into a westernised indigenous person. The content that was used to create the curriculum, the objectives of the content and the values of the curriculum are structured in a way that teachers educate from the European point of view. As a result, even indigenous languages are taught in English, from the European perspective. Undergoing for education in a westernised university as a colonised subject could mean a moment of undergoing a metamorphosis. It is a moment of morphing from being an indigenous person, a free human being whose subjectivity is not limited within any form of borders, who is not transformed into a westernised person, from being a person with free subjectivity into an object of European creation.

In history there is always a time for colonised subjects to be doing something for the first time to the point that it becomes their norm and while the historical reasons are normally forgotten or are intentionally omitted. For instance, most cultural elements that most colonised subjects protect as their identity and heritage are elements that resulted as surviving apparatus for slaves or for people in the ghettos and slums. Food, songs, stories, dances, art and language that are still adopted today by colonised subjects as elements of their cultures were created specifically for surviving under a particular kind of oppression or dispossession. In this regard, education and the curriculum are amongst the things that were created during apartheid in the case of South Africa. In this colonial set up, the colonised subject was normally allowed to acquire education that was only technical and only useful for working under the white master until post-1994, colonised people were officially allowed to take education of their choice. An educated black/colonised subject became a dangerous subject under the colonial state because they can obtain their education and use it against the system. In this regard, Fanon highlights how the French were taking over Algeria and brought newspaper for the first time. The government or the comrades against the struggle used these newspapers to communicate or to analyse what the other did or were planning to do. In the case of Algeria, Fanon writes, "the purchase of such a newspaper was thus considered to be a nationalist act. Hence it quickly became a dangerous act. Every time the Algerian asked for one of these newspapers, the kiosk dealer, who represented the occupier, would regard it as an expression of nationalism, equivalent to an act of war" (Fanon 1965, p.81). Asking for a decolonised Afrocentric curriculum in the westernised university can be considered to be an act of war. Every time a black enters at the university he/she is watched with eyes of wonder and resistance, they are accepted under the political sake, till when they leave the university, they wonder and still remain jobless. They remain suspected and unwelcome they "never succeed in gaining recognition as a colleague from the whites" (Fanon [1952] 2008, p.43). Acquiring education for the black people should be equivalent to a political act and a cry for ontological recognition in the eyes of others. However, the curriculum of the empire does not offer any form of epistemological and ontological leap to the colonised figure. Once they become educated under the westernised curriculum they seem to take the position of being a new kind of a sophisticated slave or what Malcom X call a "house Negro".

This does not dispute the important fact that there were some colonised subjects that obtained their education under the curriculum of the empire in a westernised university, whose oeuvre is relevant in today's process of decolonisation and Africanisation as well as in resistance movements. The work they produced after undergoing western training provides intellectual scaffolding and epistemic pillars for the construction of a decolonised and Afrocentric curriculum. Although their western training taught them to train and create others according to the Eurocentric model, they managed to create resistance against the curriculum of the empire, however, their resistance was limited to a certain degree.

LIMITATION OF COMPENSATORY EDUCATION SYSTEM

A response to Africa's need by Africans has resulted to the need to return to the source because, "the call for indigenous approaches to management and leadership falls within the broader cry for an African Renaissance that seeks to reclaim the aesthetics and identity of Africans (Nkomo, 2006; Makgoba, 1999; Mbeki, 1998; Mudimbe, 1988; Nzelibe, 1986)".This was also in line with the theory of Afrocentricity and decoloniality that states that Africa must reclaim itself on its own terms and conduct not according to expertise of the 'Other' (Said 2002, Spivak 1990). To reclaim itself Africa has realised its need to divorce from putting its trust into the curriculum of the empire, by eradicating the act of looking at the Western for solutions. According to Bolden (2005, p.1), "the Inter Action programme seeks to transform Africa through the development of a new generation of leaders who are encouraged not only to take up their own leadership roles but to share their insights and learning to develop and inspire others within their communities". African concepts of community, leadership and family have never promoted individualism and selfishness rather it has promoted Ubuntu. The philosophy of Ubuntu includes the standards of humanism and human solidarity, kindness, and provision to others. The spirit of Ubuntu is central to the way in which the African subject think, speak, act, and interact with other humans. Therefore, Africa with the philosophy of Ubuntu has created interest and contributed to the role of leadership and management in other places across the world.

The impact that the curriculum of the empire had on the African continent has resulted to the slow development of an Afrocentric-Africa that is informed by Ubuntu. The Africa that is present is a Eurocentric-Africa, it is produced and supported by the western logic of modernity and civilisation, it is Africa from the perspective of the West, it is the Africa of the empire. If there is any time to consider the Africa of the empire dysfunctional in its nature of management and leadership that time would be now. Ironically enough, during the time of what is known as time for freedom, democracy and equality in Africa, this dysfunctionality of the African states is not something that begun before the colonial encounter as most wrong European would claim and African's would think, it is something that consequently was derived from the European processes such as colonialism and imperialism. Was Africa dysfunctional and disorganised more after the lines at the Berlin Scramble for Africa conference were drawn to make it what it is today? The African continent is in a situation because of Western interferences. The concepts in the curriculum of the empire even today still divided Africa physically, spiritually, socially, ethnically, economically, culturally and psychologically. The essential task is to explore and discuss what appears to be African dysfunctionality in the attempt to highlight the indigenous approaches which preceded colonialism. The African continent has its own rich history and inheritance way back before the interference of the coloniser. Africa has been led and managed by great men even during the times of wars of resistance against the invasion of the colonisers. Naturally most African cultures require a great level of skill in leadership and management, these ways of living need to function in harmony to keep the indigenous unity together.

According to Nnabugwu and Nwanegbo (2012, p.131), "a major character of the contemporary Africa is the pervasive and overwhelming presence of inorganic and dysfunctional states and structures". Africa's inheritance from the colonisers made it what it is today, it made South Africa and other countries in the continent custodians of the curriculum of the empire, with alien institutions and structures that disorganised it while claiming to organise it. What the curriculum of the empire does not clearly articulate is the fact that the dysfunctional side of African states began during the invasion of the colonisers and it still carries on today. It does not state from the position of the colonised, that different states of Africa have engaged on wars, big projects of large construction and negotiations that displayed a great evidence of skill on African leadership and management even during the resistance against colonial invasion. These big historical events were used to draw solutions from African problems by Africans until the coloniser "emerged without consideration of what the people of Africa wanted or wished" (Nnabugwa & Nwanegbo 2012, p.131). However, before the colonisers came into the picture, the African states were functional in their own way although there is not much written, and research done on African management and leadership in the old days. This is also reflected in the curriculum of the empire by the fact that "within organization studies, theories of leadership and management have generally omitted the voice of the racial 'Other' whether it is Africans or other non-Western perspective" (Nkomo, 2006; Cox & Nkomo, 1990; Nkomo, 1992; Prasal, 2006). Under the curriculum of the empire, there is a great deal of danger posed by what seems to be the fundamental silencing of the non-Western narrative. Literature about approaches to management and leadership might silence the non-Western narrative and provide from anywhere in the world but it does not necessary mean there are no other approaches to management and leadership originating from non-Western concepts.

The hypocritical view that a "98 percent of leadership theory emanates from the United States and has been developed primarily by studying American leaders. Yet, leadership theory is largely represented as universal" (Nkomo, 2006, p. 3) is deceiving and shows the fact that 80% of the curriculum of the empire is Eurocentric. There cannot be a universal leadership rule because situations, places and people are not the same. However, a leader is only great for its time. This idea of taking leadership and management approaches from the West contributes to the concept of Americanising the world. It is the same as taking the curriculum of the empire as the only curriculum in the world. That is America and the rest of the West manipulates Africa to be dislocated from her own roots. The scarcity of literature that reveals management and leadership approaches in African context have been a contributing factor to Africa's delay from her own renaissance. When looking at the curriculum of the empire and its textbooks or in the history of African leadership and knowledge production, only Nelson Mandela and Shaka Zulu are found as leaders (Nkomo 2006, p.4). This is problematic because it can give the impression that these were the only leaders that were considered effective in Africa. In this subject of leadership and management Africa appears in four categories namely; African leadership management, National culture, Africa leadership, African management philosophy (Nkomo 2006). The above summary of African management development reveals that Africa needs to respond to the need of literature that contextualises her development in her own terms. The Africana subject needs to respond to the need to decolonize the curriculum of the empire and change of corrupt leadership and management. In line with this, Alavi (1972, p.61) argues, "it might be said that the 'super-structure' in the colony, is therefore 'overdeveloped' in relation to the structure in the colony, for its basis lies in the metropolitan structure itself, from which it is later separated at the time of independence…The postcolonial state inherits that overdeveloped apparatus of the state and its institutionalized practices through which the operations of the indigenous social classes are regulated and controlled". The African state is in a situation where they appear dysfunctional to the rest of the world because of the foreign leadership and management administrative tools and policies that are taught in the curriculum of the empire.

Institutional and Structural Challenges

Africa is now educating itself to fix the dysfunctional knowledge systems as stated by Thabo Mbeki that "education has an important role to play in the economic, cultural and political renaissance of our continent and in the drive for the development of indigenous knowledge systems. This implies that all education curricula should have Africa as their focus, and as a result be indigenous in its grounding and orientation. To address this state of affairs we need to distinctly Africa knowledge systems" (Mbeki 2005). The development of Africa is hinged on the participation in education of African indigenous knowledge systems and other systems of knowledge. Nkomo (2005, p.19), further argues that "the search for African leadership and management should draw upon the past but must also inevitably be rooted in the present". As much as Africa is seeking ways to develop herself from her indigenous traditions she does not do this in order to stay and remain in the past, but she uses the past in order to understand why things are the way they are in the present. To highlight the point that Africa is not dysfunctional but is organising herself to fix the problems she inherited from the West. Leadership in Africa, p. meaning, impacts and identities argues that "rather that bringing in "expects" from the "developed world" Inter Action embraces Africa wisdom, acknowledging the strength and contributions of all participates and partners, and uses a facilitative, discursive and experiential approach rather thin a more didactic teaching based format (Bolden 2005, p.1). Therefore, Africa's epistemic dislocation has become a call for Africa to find herself through the mist of colonial inheritance. Many attempts are being made by many people who believe that Africa is not where it is supposed to be because it was robbed by other people. This we can say has come to be just one of the general history lessons nothing much but for a new different African scholar like myself, the narrative about Africa is more than just a place for a picnic or something that can simply be buried under the Eurocentric capitalism carpets, it is the heart and mind of the mother earth. African narrative is told by many scholars, researchers, authors, artists and even film makers, but there is still a gap. Instead our historic stories are being told by Western movie producers and actors while our producers and actors starve. For some people the narrative about Africa is a depressing story while to others it's a story of victory, perseverance and optimism.

The depressing side of the story lies on its redundancy and contradiction of the curriculum of the empire. It has also come to be difficult to apply in Africa because government offices that should be improving people with funding and resources to apply their ideas through projects that could facilitate social change and social cohesion are corrupt and irresponsible. Most people in South Africa who are occupying these big government offices took on responsibilities that were new and new to black people. After 1994 new state departments that were not there during the apartheid government were opened and occupied by people with no experience of the job or in that field as a way of prioritising the previously disadvantage. This creates a gap between the offices and those it is supposed to serve, the community of the people. Therefore, this creates a delay in making things happen towards the Africa renaissance.

It is already known why Africa is like this, there are many books explaining this even better than the new scholars and the internet. In South Africa we have museums, statues, artworks, books, poems, tourist attractions and monuments telling the story of the colonisation and apartheid better than some black or white demagogues who are trying to use those stories for political empowerment especially when near to elections. Many people still want to fix Africa into a better Afrocentric-Africa that Africans want, but unfortunately, they still look outside themselves for solutions to their own problems. Some African leaders have lost connection with their ancestors therefore they lose African wisdom. Wisdom is an African element while enlightenment is a European element. The contradiction begins when Africans take the knowledge of the enlightenment as a solution. Current problems in Africa that are the result of resources being looted by the European powers need more than just logical and scholarly debates, but they need wisdom and political will as a radical force that we have inherited through African ancestors. What kept cultures and people united in the old Africa before it was even called Africa was the wisdom that those cultures offered.

African ancestors used African wisdom that can only be found in their different cultures to build strong and united communities until the curriculum of the empire introduced a new vocabulary to the colonised subject. This new vocabulary came with a new reality for most colonised indigenous communities. As stated, the introduction of a new word such as jail, meant that a building structure must be constructed, once its constructed there is a certain set of rules that are set to be followed by citizens in order to avoid the open-trap, jail. It is an open-trap because jails are never full, but they are full of black skin across the world. The set of rules that citizens must follow are called the law and this law is supposed to represent justice. Justice is supposed to be the big brother who is watching everyone to just to others as they would like themselves treated just. In other words, the curriculum of the empire introduced a deviated reality to the colonised subjects, it gave them vocabulary for the colonial reality of torture and lived experience that it is about to experience as the anti-black law of existing. The level of resistance against the European domination that resulted into war in Africa and even in areas like the Eastern Cape in South Africa; the war between Xhosa people, Boers, Khoi khoi, San and British which is not well spoken about or even taught as part of the South African History curriculum, shows a slow level of development. Some of these wars that happened as resistance of colonisation are not well recognised as the beginning of applicable decolonasation, although the first resistance against slave trade, was a decolonial act.

It is scientifically proven that most of our daily life is conducted under the watch of the subconscious mind, the same subconscious mind that dreams and reasons while we are sleeping, which means we do not have full control if we have any control over it. In Xhosa there is a saying that says "*imfuzo iyagqithisa*", which means offspring will live as their fathers lived because naturally there are many things that we unconsciously inherit through our parents who also have inherited them from colonial domination and their parents. These fears, insecurities, psychological dislocation, cultural dislocation and other after effects left by colonial domination are resurfacing through the curriculum of the empire creating a schizophrenic reality effect. This schizophrenic reality is what Dubois ([1903] 2016) calls, "double consciousness" as the result of such mentioned powers. The process of decoloniality is an attempt to escape this schizophrenic-reality which competes with the reality that African people are busy trying to develop for a new Africa. The schizophrenic-reality is a result of psychological and cultural dislocation and European lies that were instilled and are still being instilled in the psyche of the world for domination and capitalism benefits. Thus, Dubois explains well this effect of the curriculum of the empire in the following words:

It is a peculiar sensation, this double-consciousness, this sense of always looking at one's self through the eyes of others, of measuring one's soul by the tape of a world that looks on in amused contempt and pity. One ever feels his two-ness, an American, a Negro; two souls, two thoughts, two unreconciled strivings; two warring ideals in one dark body, whose dogged strength alone keeps it from being torn asunder. The history of the American Negro is the history o9f this strife- this longing to attain self-conscious manhood, to merge his double self into a better and truer self. In this merging he wishes neither of the older selves to be lost. He does not wish to Africanize America, for America has too much to teach the world and Africa. He wouldn't bleach his Negro blood in a flood of white Americanism, for he knows that Negro blood has a message for the world. He simply wishes to make it possible for a man to be both a Negro and an American without being cursed and spit upon by his fellows, without having the doors of opportunity closed roughly in his face ([1903] 201, p. 62-3).

Unfortunately, Africa has been a victim and still is of these colonial forces that are pushed for capitalist benefits of the empire. There is an African proverb that says, "if the is no enemy within the enemy outside can do us no harm". The curriculum of the empire has created black leadership and management in Africa that is redundancy and contradictions in the African narrative or history has created enemies within Africans. It has created divisions in Africa with epistemic borders, geographical borders, global borders, social and political borders and ontological borders.

POSSIBILITIES OF A TRANFORMED CURRICULUM: A DECOLONIAL AFROCENTRIC PERSPECTIVE

Blackness and the curriculum of the empire offers "an existential deviation on the Negro" (Fanon 1952[2008], p. 6), as a position that encourages one to question his or her own existence outside the curriculum. Blackness is the place of the forgotten, a place of the dispossessed and a place of the figure, the one who can be anything, the one with many faces, a place of nonexistence. Blackness and the curriculum of the empire offers two separate realities that are positioned to work together, this combination is a colonial tactic of giving the colonised subjects education that does not speak to their identity, experience and history. It is from the position of the "Afrocentric theory" (Asante 2007, p.10; Bangura 2012, p.110) that the question of the curriculum of the empire reveals that the colonised subjects are not included under the umbrella of nationalism, because as stated by Fanon "history teaches us clearly that the battle against colonialism does not run straight away along the lines of nationalism. For a very long time the native devotes his energies to ending certain definite abuses, p. forced labor, corporal punishment, inequality of salaries, limitation of political rights, etc" (Fanon [1961] 2001, p.119). With Fanon's remark, it is clear some people seem to remain excluded as the idea of the new-nationalism is questionable. This exclusion is manifested in different ways such as being excluded in the curriculum, being excluded in the library archives, being excluded in the art gallery and museums, being excluded from the human register.

The curriculum of the empire is modern thinking because, "modern Western thinking is an abyssal thinking. It consists of a system of visibility and invisibility, the invisible ones being the foundation of the visible ones… The division is such that "the other side of the line" (Santos 2007, p.33) vanishes as reality becomes non-existent, and is indeed produced as nonexistence. Nonexistence means not existing in any relevant or comprehensible way of being", hence the curriculum of the empire as well as modernity are contingent upon the exclusion of the colonised people. Their existence was fueled by nonexistence of the colonised subject because they are both designed and constructed in a way that keeps the colonised subject different and constantly under the state of seeking. Seeking education, seeking a better life, seeking recognition as a human, seeking identity and seeking language to articulate their existence that is always questioned. In this regard Fanon argues ([1961] 2001, p.119), "national consciousness, instead of being the all-embracing crystallization of the innermost hopes of the whole people, instead of being the immediate and most obvious result of the mobilization of the people, will be in any case only an empty shell, a crude and fragile travesty of what it might have been". Meditating deeply on Fanon's remark, it becomes clear in the case of South Africa, that not all South Africans are South Africans, which is to say, not everyone is included under the national identity. This brings under question the notion of a rainbow nation that was coined and introduced by Nelson Mandela as an alternative solution to apartheid's social and political ills. Under the notion of a rainbow nation, what Mandela was stating was that after the fall of the apartheid system African/ black people of South Africa must not do what the coloniser did to them, they must consider a more inclusive and assimilative approach.

The rainbow nation as an assimilative approach of being inclusive is a process of becoming, because to assimilate is to become someone or somethings else. In this process, Mandela was arguing that colonised/black people do not have to do what the white people did to them, in other words, in the quest for freedom and justice, the colonised/black figure will not take the position of being the oppressor because the "curriculum was biased toward European knowledge and Western styles of teaching, and that this bias was harmful to the self-esteem and performance" (Binder 2002, p. 1) of the colonised figure. The consciousness of the rainbow nation was limiting as it considers colonised/black people human only to a certain degree, and this degree is the one that is set by the curriculum of the empire. However, the fundamental base of the curriculum of the empire is that it is Eurocentric, and it introduces a version of a Eurocentric-Africa.

Indigenous Knowledge Systems: An Introduction to Afrocentric Perspective

According to Smith (1999, p. 138), "indigenous research approaches problematize the insider model in different ways because there are multiple ways of being either an insider or an outsider in indigenous contexts" (Smith 1999, p. 138). The problem that is faced by the world is the problem of a single side allegory of how Europe is the father of knowledge, being and power amongst everything in the world. When conducting research about the colonised people, it is important to deploy a theoretical framework and methodology that positions colonised/black people as a centre of their own world. Although there are many research tools that can be deployed by colonised/black people to conduct research about their existence, very few or none of them explain the phenomenon from their position as the centre. Afrocentricity as a theoretical and philosophical perspective offers decolonial "actionable perspective" (Asante 2007, p.16). Often the story glorifies the hunter while the hunted is always misrepresented, in this case, Europe is always represented as the centre while other countries are misrepresented as inferior. Therefore, in order to relocate the colonised/black people in general, the African people in particular, it is important to deploy decolonial epistemic perspective as a theoretical framework and Afrocentricity as a methodology. According to Binder (2002, p. 30-31):

The Afrocentric methodology rests on two primary assumptions. The first principle is that analysis of any subject with roots in Africa must begin with "the primacy of the classical African civilizations, namely Kemet (Egypt), Nubia, Axum, and Meroe. . . Adequate understanding of African phenomena," according to Asante, "cannot occur without a reference point in the classic and most documented African culture... The second major premise upon which Afrocentrism is based is that all people of African descent possess essential cognitive, cultural, and aesthetic characteristics in common. The wife of Molefi Asante, an Afrocentric scholar in her own right, KariamuWelsh-Asante, for example, argues that there is an African Aesthetic, which is "based on seven 'senses' shared by all Africa-descended people around the globe, p. polyrhythm, polycentrism, dimensional, repetition, curvilinear, epic memory, and wholism".

In this regard, Afrocentricity is a suitable methodology because it deals with the colonised/black figure the body and mind level, decoloniality deals with the soul. For most indigenous societies, the soul encompasses essence and essence is being and existence. Therefore, it is important to engage a methodology that poses existential questions as the insider or outsider. The research methodology that is proposed here is still underdevelopment, which is the decolonial Africana existential phenomenology. Phenomenology study is a subjective experienced based research methodology that received a lot of attention in the mid-21st century in the field of qualitative research along with autoethnography because it was regarded as a "process whereby we encounter ambiguity in texts and in all cultural products. We find in the very ambiguity new possibilities for thoughts and actions," (Lundin et al 1985, p. 32). According to Langdridge phenomenology is a study that "aims to focus on people's perceptions of the world in which they live in and what it means to them; a focus on people's lived experience" (Langdridge 2007, p. 4). Phenomenology was popularized in the early 20th century to affirm abstract "realities" (Gable and

Handler 2003, p. 376) "lived experience" (Ingold 2000, p. 145) and real life "events" (Zizek 2014, p. 3), as well as the feelings and actions of people. Phenomenology was used by a phenomenologist Maurice Merleau-Ponty's in an essay Cezanne's Doubt, where he "was particularly interested on how Cezanne saw and related to the world and created his paintings the way he did. More especially he was interested to the relationship between the eye and the hand in relation to the creation of the paintings" (Kozel 2013). Phenomenology as a methodology has been applied in various fields of studies including, nursing, biology, psychology, performance arts, therapeutic art and other studies that focus on lived experience. Phenomenology is divided into transcendental, hermeneutic, and existential phenomenology. In most cases existential phenomenology is deployed because it "stresses on the description of everyday experience as it is perceived by the consciousness of the individuals" (Kafle 2011, p.188). In this regard, decolonial Africana existential phenomenological as a methodology draws from existential phenomenology and it is based on the need to the return to lived experience of the colonised people and their consciousness of the world they leave in. Decolonial Africana existential phenomenological is a "research methodology aimed at producing rich textual descriptions of the experiencing of selected phenomena in the life world of individuals that are able to connect with the experience of all of us collectively" (Smith 1997, p.80), specifically black colonised people's phenomenon. Decolonial Africana existential phenomenological study is deployed to affirm theory and practice within the academic research and allows the embodied lived experiences to provide new perspectives. Decolonial Africana existential phenomenology is, "a "gathering" notion under which to situate the articulations (writings, speeches etc.) and traditions of the same, of Africans and peoples of African descendent collectively, as well as sub-discipline—or field-forming, tradition-defining, tradition—organizing reconstructive efforts which are (to be) regarded as philosophy" (Outlaw1997, p.267). Decolonial Africana existential phenomenology as a methodology will be deployed as a method of this study to reflect on "lived experience" (Ingold 2000, p.145) in critical academic setting. Existential phenomenology was deployed by philosophers namely; Edmund Husserl, Martin Heidegger, Maurice Merleau-Ponty, Hans-Georg Gadamer and Paul Ricoeur as a method to reflect on Dasein, which means to 'dwell in the world' or "being-in-the-world" (Heidegger 1962). Decolonial Africana existential phenomenology, in a similar way with African existential phenomenology, according to More is based on "reflections, rooted in black experience, on the boundaries of human existence and the utilization of such reflections to challenges confronting African and African-descended people in Diaspora" (More 2008, p.47). Decolonial Africana existential phenomenological is a critical and creative methodology of no method that resonates well with the intension of this study, which is to diagnose, prescribe and make dead objects come to life and speak or resist. This research is deployed to interpreted black people's lived experience as reflected in their art to understand the culture and being of the black people because "ancient and modern tribal peoples would not distinguish art from artefact or ritual" (Freeland 2013, p. xviii). As it has been proven by various scholars, certain methods prove to carry colonial tools that do not justify the phenomenon of black people. Decolonial Africana existential phenomenological is a research methodology whose intension is to dress and conceal colonised people who have been stripped naked by dirty colonial tricks of research. It is a method to reflect on the notion of being-in-the-anti-black-world.

Decolonial existential phenomenological is a way of turning lived experiences into a critical academic mode and challenge the westernised ways of living, writing, art making and ways of seeing. Decolonial Africana existential phenomenology is a platform to use lived experiences of dehumanised people to formulate counter points of the historic gap created by western domination. However, according to Outlaw, "in the context of such endeavours, persons of the past and present without formal training or degrees in philosophy are being worked into developing canons" (Outlaw 1997, p.266). Decolonial Africana existential phenomenology does not mean the experience of maids, including my experience as a black person; and Fanon's are to be embraced as absolute truth, but it embraces the position that any person's embodied experience when properly contextualised and reflected upon offers perspectives and meaning that speak to other people's realities. Decolonial Africana existential phenomenology seeks to move away from aesthetics to aesthesis (Mignolo and Vazquez 2013). In line with aesthesis, More asserts that "by virtue of the historical fact of racial oppression, colonisation, and … [bondage], Africana philosophy [and art] raises questions of identity and liberation by focusing on the reality that black …[figures] are affected by the significance of race and racism" (More 2008, p. 47). The problem with aesthetics according to More's remark on aesthesis, it looks and sees the world through the eyes of race and racism. Meaning through the eyes of race and racism everything else collapses into nonexistence including ontology, law, beauty, reason, history and art when it comes to the black figure. According to Seshadri-Crooks the "scientific basis of race has been thoroughly discredited" and that race is "purely cultural, a construct, and the racialised body would be the imaginary corollary to that construction" (Crooks 2000, p. 12-21). Decolonial Africana existential phenomenology therefore, similar to Africana existential philosophy, as according to More, "consists in reflections, rooted in black experience, on the boundaries of human existence and the utilization of such reflections to challenges confronting African and African- descended people in Diaspora" (More 2008, p.47). One big challenge confronting black colonised people is the challenge of being able to express and articulate the colonial mistreatment in all its subliminal inscription. It is the challenge of lacking tools of excavating coloniality and realise it is not buried underground rather it is surrounding us.

Regardless of what appears as an inability to articulate black people's existence as human beings, Fanonian decolonial phenomenological study "deals with issues of the emergence of the black selfhood, black suffering, embodied agency, freedom, bad faith, racism, and liberation" (More 2008, p.47). By deploying Decolonial Africana existential phenomenology from the positionality of aesthesis in order to make sense and meaning of being-in-the-anti-black-world, Diagne argued that it must deploy what Bergson teaches us on how to make sense about philosopher's thoughts. In regard to making sense with aesthesis, Diagne argues that, "we must begin by taking a step back from …[its] thought in order to return to its source, weigh the influences that nurtured it. And pinpoint the ideas of which the doctrine is a synthesis" (Diagne 2011, p.00). Diagne's approach to reading philosophers and aesthesis can be linked and viewed in line with decolonial Africana existential phenomenology to read the colonised/black people as a manifestation of a colonial phenomenon from the positionality of returning to its source. From the decolonial Africana existential phenomenology point of view, the difference between a Eurocentric and Afrocentric is how each introduces the world to the viewer by "… looking, seeing and representing visually with the problematic of [racial] difference" (Pollock 1988, p. xvii). This means making sense of the world by deploying all senses of the body to generate knowledge and meaning. It means not only relying only on thinking or seeing as the Western canon has suggested.

Decolonial Africana existential phenomenology makes sense of the world by moving beyond the surface appearance to the deeper essence of the live experience. Hence, "Afrocentrism recognizes no division between the African past and African American history, and regards as ahistorical and mythical any social science that does not trace these continuous African Roots" (Binder 2002, p. 31). In a similar way to the Afrocentrism, decolonial Africana existential phenomenology "…must begin by retrieving the initial attitude, the hermeneutic posture that Senghor adopted from his early writings onwards in order to answer the question (which, as we will see, was also Picasso's), p. What do African mask mean?" (Diagne 2011, p. 9). Decolonial Africana existential phenomenology comes from a similar position posing a similar question that is connected to the African mask whose meaning is undefined from a colonial gaze, p. what does it mean to exist as a black person in the anti-black world? Decolonial Africana existential phenomenology as a method is a way to move beyond the surface of appearance and get to the essence in attempt to reflect on existential and ontological questions. These questions according to Diagne, can only be answered if, and "when we go back to its source and weigh the influences to pinpoint the ideas of which it is a synthesis…" (Diagne 2011, p.4). Therefore, decolonial Africana existential phenomenology enables objects to think, speak and resist in order to find and embrace the sublime in the grotesque. It "correct[s]… history books for their neglect or oversights of African experience in one of two ways, p. by asserting the preeminence of Africa and its descendants in the course of world historical events, or by underscoring the presence of subjugated African peoples throughout time and the remarkable triumphs of those oppressed over systems of oppression" (Binder 2002, p.31). That is, finding life in being-in-the-anti-black-world and making sense of the lived black experience.

CONCLUSION

In concluding on something that is difficult to conclude about, it is clear that indigenous knowledge systems will continue to suffer as long as the current Eurocentric logic is prevailing, the current anti-black world still stands, and Afrocentricity is absent in the curriculum. This means the is a space open for future research on the topic of decolonising the curriculum by injecting indigenous knowledge systems. The curriculum of the empire is supported by whiteness and white supremacy that positions itself as the defining factor of what is the standard of learning and what is learnt, the standard of knowledge, being, power, beauty and life itself. Decolonisation of the curriculum of the empire means discrediting it from the Afrocentric position by injecting or replacing it with indigenous knowledge systems. It becomes clear that indigenous communities have a serious task at hand, which is to rethink and decolonise their indigenous knowledge systems. Indigenous knowledge systems suffered under western episteme because of its inability to fully articulate the lived indigenous experience, instead of dehumanizing it. This calls for new decolonial existential and phenomenological approaches that are set to recontextualise, reconfigure, rethink and re-membering of the African subject as a human being. In other words, there is a deep need for the decolonisation of the curriculum and repositioning of indigenous knowledge systems as centre knowledge for the indigenous people in the trans-modern world non-Manichean world. This means the remobilisation of African/black/indigenous thinkers and activist who challenged the empire while still located in blackness by deploying indigenous methods and philosophies. Decolonisation/deoloniality is a project that is lead across the world, it is a project for all the colonised subject who are resisting modernity/coloniality under the umbrella of blackness. Therefore, it is important to remember for the colonised subject who engages in the process of decolonisation that, decolonisation/deconiality is only a bridge to help the colonised subject to cross the river of modernity/coloniality and arrive at a particular destination. Hence decolonisation/decoloniality is applied in this article in relation to the Afrocentric perspective, the aim is to epistemologically and ontological arrive to a decolonial Afrocentric curriculum.

REFERENCES

Asante, M. K. (2007). *An Afrocentric Manifesto, p. Toward an African renaissance. Polity.*

Bangura, A. K. (2012). From Diop to Asante, p. Conceptualizing and Contextualizing the Afrocentric Paradigm. *The Journal of Pan African Studies*, *5*(1), 103–125.

Binder, A. J. (Ed.). (2002). Contentious curricula. In Afrocentrism and creationism in American public schools. Princeton University Press.

Ce'saire, A. (2000). Discourse on Colonialism (J. Pinkham, Trans.). New York: Monthly Review Press.

Cox, T., & Nkomo, S. M. (1990). Invisible men and women, p. A status report on race as a variable in organisation behaviour research. *Journal of Organizational Behavior*, *11*(6), 419–432. doi:10.1002/job.4030110604

Du Bois, W.E.B. (2016). *The Souls of the black folk.* Digireads.com Publishing. (Original publication 1903)

Fanon, F. (1965). A dying colonialism (H. Chevalier, Trans.). New York: Monthly Review Press.

Fanon, F. (1965). A dying colonialism (H. Chevalier, Trans.). New York: Monthly Review Press.

Fanon, F. (2001). The wretched of the earth (C. Farrington, Trans.). London: Penguin. (Original publication 1961)

Fanon, F. (2008). Black skin, white masks (C.L. Markmann, Trans.). London: Pluto. (Original publication 1952)

Ghandi, L. (1998). Postcolonial theory. In A Critical Introduction. Edinburgh University Press.

Gordon, L. R. (1995). *Bad faith and anti-black racism.* Humanities Press.

Gordon, L. R. (2006). *Disciplinary decadence. Living thought in trying times.* Paradigm.

Hardt, M., & Negri, A. (2000). Empire. Harvard University Press.

Lorde, A. (1984). The Master's Tools Will Never Dismantle the Master's House. In Sister Outsider. Crossing Press.

Makgoba, W. (Ed.). (1999). African renaissance. Mafube Publishing.

Mbeki, T. (1998). *Africa: The time has come*. Tafelberg.

Mbeki, T. (2005). *Goals of Higher Education in Africa*. Retrieved from www.utexas.edu/conferences/africa

Mohanty, C. (1984). Under western eyes: Feminist scholarship and colonial discourses. *Boundary, 12*(3).

Mudimbe, V. Y. (1988). *The invention of Africa: Gnosis, Philosophy, and the order of knowledge*. Indiana University Press.

Ndlovu-Gatsheni, S. J. (2013a). Empire, global coloniality, and Arican objection. Berghahn Books.

Nkomo, S. (1992). The emperor has no clothes, p. Rewriting 'race into the study of organizations'. *Academy of Management Review, 17*(3), 487–513.

Nkomo, S. M. (2006). *Images of African Leadership and Management in Organisation Studies: Tensions, Contradictions and Re-visions*. University of South Africa.

Nnabugwu, M.B., & Nwanegbo, J. (2012). Dysfunctional state and severance in Africa, p. an analysis in specificities and contradictions. *Ogirisi: A New Journal of African Studies*, *9*, 127-140.

Nzelibe, C. O. (1986). The evolution of African management thought. *International Studies of Management & Organization, 16*(2), 6–16. doi:10.1080/00208825.1986.11656427

Owusu-Ansah, F. E., & Mji, G. (2013). African indigenous knowledge and research. *African Journal of Disability, 2*(1), 1–5. doi:10.4102/ajod.v2i1.30 PMID:28729984

Oyewumi, O. (1997). The invention of women, p. making an African sense of Western gender discourse. University of Minnesota Press.

Santos, B. S. (2007). Beyond abyssal thinking from global lines to ecology of knowledges. *RE:view, 30*(1), 1–33.

Smith, L. T. (1999). Decolonising methodologies: Research and indigenous people. Zed Books.

Spivak, G. C. (1990). The post-colonial critic. Viking.

Thiong'o. (1993). *Moving the centre: The struggle for cultural freedoms*. East African Educational Publishers Ltd.

Yountae, A. (2017). The decolonial abyss: Mysticism and cosmopolitics from the ruins. Fordham University Press.

KEY TERMS AND DEFINITIONS

Africana Existential Phenomenology: It is a kind of phenomenology that allows Africana people to study their subjectivity in relation to the colonial world based on their lived experience. By putting the word decoloniality as a prefix suggest that it is located in the project of decolonisation.

Afrocentricity: An approach that positions African people as the centre of their knowledge production. It allows the African subject to engage the world based on their indigenous knowledge systems.

Blackness: It is a kind of a condition that is created through racism, imperialism and colonisation. It used here from a decolonial epistemic perspective, which believe blackness is a cause and fuel of modernity.

Coloniality of Knowledge: It is a form of colonisation that privileges Eurocentric knowledge over other knowledge systems. It labels indigenous knowledge systems as primitive knowledge that is not at the level of European knowledge.

Decoloniality: It as project of challenging colonisation and racism. Unlike post-coloniality which is located inside modernity, decoloniality comes outside of modernity.

Empire: It is a concept that was invented by Michael Hardt and Antonio Negri (2000) to suggest the global political world that is ruled as one empire, which is controlled by a certain group of people. In the empire, it is only the white subject who is considered a sovereign subject who also qualifies to be the only human.

Frantz Fanon: Is a Black theorist who is known for his work which focuses on the lived experience of the black subject in the colonised world or empire. His work is a critique of colonisation and racism, it allows the black colonised subject to be able to understand their subjectivity in a modern/colonial world.

Indigenous Knowledge Systems: It is a knowledge system that predates colonisation and European modernity. It is an ancient knowledge that is based on the way indigenous people see, think and do in the world based on their culture.

Section 2

Discovery of Indigenous Moral Standards in the Learning and Research Methodology

Chapter 4
The Significance of Multicultural Methodologies on African Indigenous Knowledge Research

Stewart Lee Kugara
University of Venda, South Africa

Tsetselelani Decide Mdhluli
University of Limpopo, South Africa

Thizwilondi Joanbeth Daswa
University of Venda, South Africa

Pfarelo Eva Matshidze
University of Venda, South Africa

Ndidzulafhi Esther Ramavhunga
University of Venda, South Africa

ABSTRACT

Throughout history, African indigenous knowledge has been researched using Western-centred methodologies. From the onset, there was a great deal of resistance from indigenous peoples as most of these researchers used ethical considerations that are foreign to local communities. The African indigenous holders emphasise some cultural concepts that are sacrosanct and indispensable to them. This chapter, therefore, seeks to ferrate the significance of Western-centred methodologies vis-à-vis African-centred (Africa-centred, Afrocentric, Africentric, Afrocentricity, and African-worldview) methodologies in the collection of data among African indigenous communities. The concept of sharing (hybrid methodology) provides a platform of reflection especially on protocol and research tools.

DOI: 10.4018/978-1-7998-1249-4.ch004

INTRODUCTION

For some time, African indigenous knowledge has been researched using Western-centred methodologies. From the onset, there was a great deal of resistance from indigenous peoples as most of these researchers used ethical considerations that are foreign to local communities. The African indigenous holders emphasise some cultural concepts that are sacrosanct and indispensable to them. Resultantly, if such are not adhered to, this result and/or has resulted in numerous African indigenous peoples not disclosing the information or giving half-truths thus, the researchers producing findings that are a mismatch from reality. This is so as the ethical strategies and methodologies adopted are processed from a purely Western perspective, without respecting the community of origin. This proposed book chapter, therefore, seek to ferrate the significance of Western-centred methodologies vis-à-vis African-centred (and its variations as Africa-centred, Afrocentric, Africentric, Afrocentricity, Africentricity, and African-worldview) methodologies in the collection of data among African indigenous communities.

It is trite that researchers should do a critical analysis of African indigenous peoples' life, history and culture from the perspective of African indigenous people. In that regard, 'some forms of behaviour, actions and conduct are approved while others are widely disapproved of' (Idang 2009, p. 142). In evaluating African culture and qualities, the book chapter is not pontificating that every African culture have the equivalent explanation(s) for occasions, a similar language, and same method of dressing, etcetera. Relatively, there are fundamental likenesses and ethics shared by numerous African indigenous communities which if not adhered to introduce a mismatch and bring an aspect of undermining. If such cultural practices are disregarded, African indigenous communities take on responsibilities and distort or frustrate what the researcher seek.

On that backdrop, using Western-centred ethics on African indigenous knowledge research is not only dehumanising but portray them as 'pawns'. As such, the book chapter intends to bring this novelty, correct the wrong and introduce an adaptive system which upholds the values that play a central role in African indigenous communities. The book chapter concluded that it is important that the framework or methodology that is employed to research African phenomenon is grounded in the African Indigenous knowledge systems paradigm, because the community of research should be respected. Worth-noting, these two systems, Euro-Western centric and African methodologies can co-exist thus need to be respectfully compared and contrasted without restraint.

The concept of sharing (hybrid methodology) provides a platform of reflection especially on protocol and research tools. The adoption of Euro-Western methodologies with the exclusion of Afro-centred methodologies in sharing platforms is alleged to be toxic. It is toxic because it has a tendency of persistently altering the identities that are known and misses the whole essence of the research purpose. As a basic principle, researchers (especially 'outsiders') who come to African indigenous communities should be vigilant on how to design projects so as to safeguard the dignity and equity of their participants. If this is not taken seriously, it has potential of unveiling hard questions about how a researcher's positionality has causal effects. Thus, guidelines of important ethics to note in African protocols and values were ferreted. The chapter critically assess who are African indigenous peoples, their indigenous knowledge, appraise Euro-Western methodologies and African ethics and protocols.

African Indigenous Peoples

Definitions and Conceptualizations of Indigenous Peoples

"There are many indigenous peoples globally and because of the diversity of indigenous peoples there is no official or universal definition of indigenous people. However, most writers use it to refer to first peoples who are minorities in their own lands. The terms "indigenous peoples," "indigenous ethnic minorities," "tribal groups," "Aboriginal", "Native", and "First Nations" and "scheduled tribes" describe social groups with a social and cultural identity distinct from the dominant society that makes them vulnerable to being disadvantaged in the development process."

The International and Regional Framework for the Identification of Indigenous Peoples

"The international community has not adopted a definition of indigenous peoples. In fact, the position of most international bodies charged with examining or addressing the rights of indigenous peoples (Including the position underlying existing international legal instruments, such as International Labour Organisation (ILO) Convention No. 169 that specifically protects the rights of these peoples, as well as the position of the African Commission on Human and Peoples' Rights) is that a strict definition of indigenous peoples is neither necessary nor desirable."

It is much more relevant and constructive to try to outline major characteristics which may help us to identify who are the indigenous peoples and communities in Africa. In addition, the lack of definition should by no means constitute an obstacle to addressing the substantial issues affecting indigenous peoples.

Characteristics to Identify to Indigenous Peoples and Communities in Africa

Indigenous peoples can be identified in particular geographical areas by the presence in varying degrees of the following characteristics:

- Indigenous peoples are socially, culturally and economically distinct
- Their cultures and ways of life differ considerably from the dominant society and their cultures are often under threat, in some cases to the extent of extinction.
- With a close attachment to ancestral territories and to the natural resources in these areas;
- They suffer discrimination as they are regarded as 'less developed' and 'less advanced' than other more dominant sectors of society;
- They often live in inaccessible regions, often geographically isolated and are subjected to various forms of marginalisation, both politically and socially;
- They are subject to domination and exploitation within national political and economic structures that are commonly designed to reflect the interests and activities of the national majority.
- Self-identification and identification by others as members of a distinct cultural group.

Legal and Constitutional Recognition of Indigenous Peoples in Africa

"At the outset, it is clear that in Africa there is very little formal constitutional or legislative recognition of indigenous peoples as such in legislation. However, in practice, in some cases, these groups are referred to by their ethnonyms in many countries, and often there may be an implicit acceptance that certain groups within the state boundaries identify themselves as indigenous peoples in accordance with the international understanding of the term.

A number of states, such as Eritrea and Ethiopia, officially recognise the different ethnic groups residing within the state boundaries. In Ethiopia, for example, the 'nations, nationalities and peoples' within the state are officially recognised in the Constitution and accorded specific constitutional rights'. But often, it is the case that such formal recognition does not translate into concrete implementation of rights in accordance with the differing needs of different groups."

On the other hand, a number of States' constitutions and legislation do not reflect the ethnic composition of the country. For example, for Egypt, whose Constitution refers to the Egyptian peoples as an Arab nation thus, indicating that it sees itself as homogenous, failing to officially acknowledge the existence of groups that are culturally distinct from the Arab population. It is only in the equality clause of the Constitution that the very existence of linguistic, ethnic or religious differences among Egyptians is even implicated.

"Some constitutions expressly forbid any reference to ethnicity at all. Such is the case with the Rwandan Constitution. No legal text in Rwanda uses the term 'indigenous', and the government does not recognise the term 'indigenous peoples'. However, the Batwa of Rwanda and the 'Pygmies' of central Africa more broadly, are generally recognised as the first inhabitants of the region. The question of ethnicity, especially in Rwanda, is extremely sensitive and politicised and the state does not recognise the Batwa based on the reasoning that this may lead to tribalism. Rather, Batwa are designated officially as 'historically marginalised communities.'"

Indigenous Peoples in South Africa

Though South Africans have always been aware that the San were the aboriginal population of the region, it has only been since the emergence of a democratic form of government during the 1990s that groups of people have started to lay claim to this status. Notably, aboriginal San form only a small percentage of those currently claiming the status of indigenous peoples. The popularity of claiming an indigenous identity has been influenced by a number of factors, including the increased awareness of the UN's Decade of the World's Indigenous People and the emergence of a more open and democratic society in South Africa whereby previously marginalized voices are being heard.

"There is at present, no accepted South African norm as to the meaning of indigenous. The term appears twice in the Constitution (section 6 and 26), Act 108 of 1996. Familiarity with South African political discourse suggests that 'indigenous" as it is used in the Constitution refers to the languages and legal customs of majority Bantu-language speakers in contrast to those of the minority European settler population. This meaning is the norm in Southern Africa where both Botswana and Zimbabwe regularly use the term indigenous to distinguish the black majority from the European and Asian settler minorities."

Up until 1998, Khoi and San peoples did not have a place in South African political discourse. Under apartheid, the system of legalized racial discrimination, Khoe and San people were invisible, being forced to accept the racial category of coloured. The negotiations between the Department of Constitutional Development (DCD) and Griqua, Nama and San communities suggests that the South African Government is willing to review the legal meaning of the term indigenous to fit more precisely with UN and international labour Organization (ILO) usage.

"The South African Government's caution about recognizing the UN definition is not as a result of a prejudice against aboriginal populations, as is common in other parts of Africa and the world. Rather, there is a sensitivity about triggering separatist movements by larger constituencies and promoting the principle of ethnic territoriality in a State that has just escaped the tyranny of minority apartheid rule that exploited that very principle. There may well emerge two parallel definitions of the term indigenous;"

- The one broadly referring to all South Africans of African ancestry,
- The other referring to specific populations using the UN definition and making specific territorial and cultural claims against the State.

Outline of Terms

Legal and Constitutional Recognition of Indigenous Peoples

Other forms of recognition in constitutions and national legislation may be through the recognition of indigenous languages. In South Africa, the Constitution makes express mention of indigenous languages. Section 6(2) of South African Constitution obliges the state to take practical and positive measures to elevate the status and advance the use of indigenous languages. Section 6(5) of the Constitution promotes the protection of Khoi, Nama and San languages. While the three (3) indigenous languages are not given the same official status as other predominantly Bantu African languages, the fact that they are mentioned in the Constitution signals that the State acknowledges their importance. Section 6(2) of the South African Constitution refers to 'recognising the historically diminished status of the indigenous languages of our people.'

National Policies and Programmes

In South Africa, 'the cabinet adopted a memorandum in 2004 setting out a policy process to recognise Khoe and San as vulnerable indigenous people communities. This is a positive step towards the recognition of indigenous peoples in South Africa, and as illustrated by a variety of government and official statements, the Khoe and the San are regarded by the State as being an 'indigenous people' who have been marginalised and deserve special protection. However, the memorandum and public officials' statements have not been translated into an official policy recognising the Khoe and the San as the indigenous peoples of South Africa.

Indigenous (Customary) Law

Indigenous law in South Africa is also called African law (formerly also Bantu law). In other words, this was a law (culture) of black indigenous people. In is worthy to note that customary law is termed African indigenous law by modern authors. The Black Administration Act 38 of 1927 was the first law to recognise customary law. Customary law is not a primary source of law as it only apply to black indigenous.

Possibility of Customary Law

1. Applied in cases where both parties choose to have their dispute regulated by it - It will not apply automatically
2. Both parties to the dispute must be indigenous **black persons**
3. They must have maintained a traditional way of life, and
4. Respect the customary law in force in their area / their particular community

In this regard, it could then be deduced that indigenous peoples are a group of people that are governed by indigenous law. With this line of reasoning, South African black people are categorised as 'indigenous peoples'. However, several debates arose on the categorisation of indigenous peoples are black.

Black Is a Generic Term

The term "black people" is a remnant of the apartheid regime and thus regarded as offensive. Nevertheless, legislation such as the Broad-based Black Economic Empowerment Act uses the term to distinguish between people to whom the Act applies and to those to whom it does not apply. In terms of section 1 of this Act, the term "black people" is generic term that includes "Africans, Coloureds and Indians". This definition has been extended by the High Court to include Chinese people. As a result, it would be no longer be terminologically correct to use the term "black people" to distinguish the group of people to whom customary law applies. "Indigenous peoples" is used in recent legislation dealing with customary-law issues; Customary Marriages Act – "indigenous African peoples of South Africa". Although this term is also not used without its complications within the context of the Act it is accepted that "indigenous African peoples of South Africa" refers to South African ethnic groups traditionally grouped together on the basis of language and cultural features.

African Indigenous Knowledge

Basis of Indigenous Knowledge

To understand indigenous knowledge, it is essential to have a contextual background of the African society and the cultural features that are important to the beliefs of their values and norms. Prior colonialism indigenous people had their own knowledge which they passed from generation to generation, by word of mouth and cultural ceremonies. This knowledge has been the basis for educating their children, food preparation, agriculture, health care that sustain their societies and their environment. Colonialism posed a serious threat to indigenous knowledge. The colonizer mind view indigenous knowledge as primitive and perceived it as wasteland of non-achievement. The cultural wrongness historically covered both the practice and the structure of this knowledge.

The custodians of this tradition and customs in different African communities are the elder's traditional leaders. According to Sefa Dei *et al* (2000, p. 116), indigenous knowledge is a body of knowledge associated with the long-term occupancy of a certain place. This knowledge refers to traditional norms and social values, as well as to mental constructs that guide, organise, and regulate peoples' way of living and making sense of their world. This knowledge in the tradition and customs is transferred from generation to generation using the word of mouth. For instance, oral histories, storytelling and folklores. It is therefore clear that Indigenous Knowledge Systems refer to the special knowledge that specific communities have which has helped them meet their specific day by day requirements for survival (Warren, 1991 & Flavier *et al* 1995). This is the original native or home-grown knowledge which others might refer to as local knowledge.

Scholars define indigenous knowledge (IK) differently. According to Matsika (2012, pp. 209-210), indigenous knowledge is the local or traditional knowledge that existed and developed through the practises of the local communities in the process of handling the environments that challenge the community's everyday life. Woolman (2001) defined indigenous knowledge as a comprehensive performance set of events, beliefs, values and practices that has developed over time and is practiced among communities. Boateng (1983) stated that indigenous knowledge is the education of the Africans before the coming of the Europeans. It is an informal education that prepared Africans for their responsibilities as adults in their communities. On the other hand, Mushi (2009) defines African indigenous education as a process of passing down norms and values, the inherited knowledge, cultural traditions of the society and skills from one generation to another. On the roots of the definitions above, it is easy to understand that indigenous education was tangible, effective, positive and clearly comprehensible.

In other words, indigenous knowledge refers to ways of learning and teaching in Africa society which are based on indigenous knowledge gathered by Africans over eras. This knowledge is formed and transmitted by society over time, with the aim to cope in their different ecological, socio-cultural, agricultural, political and physical challenges. The spirit of the African worldview revolves around cultural values, which are comprised of wholeness, community and harmony. So even its methodology should be based on African ways of life. However, Merriam (2007) argued with the existence of indigenous knowledge. The author mentioned that indigenous education exists only in the mind of the people. Scholars such as Lander (2000) and Chavunduka (1995) have disputed that western worldview of "indigenous knowledge" lacked an appreciating of the holistic nature of knowing and in parting knowledge. Nkondo (2012) argues that the western view of indigenous knowledge as ordinary imitation of practices without any model to explain them is a portrayal of western world view.

Nevertheless, oral tradition which is characterized of indigenous knowledge systems is often looked down upon compared to the written tradition. Partly because indigenous knowledge was not documented, instead it was passed orally (Emeagwali, 2003). Though it was looked down upon, it possess rich complexities that can be observed in the community ceremonies such as rituals, initiation schools, recitation, demonstration, riddles, story-telling, poetry, word games, proverbs, folktales, songs, traditional dance, music, and other education-centred activities (Ngara, 2007).

Purposes of Indigenous Knowledge

Even though indigenous knowledge differs from one society to another, the aims of these knowledge were similar. Sifuna and Otiende (2006) suggested that indigenous knowledge was basically an education for living whose main purpose was to train the growing children to be responsible adults within society. The emphasis was to instil cultural norms and values so that growing children live the accepted standards of their community. According to Fafunwa (1995), indigenous knowledge is holistic in nature and aims at preparing growing child to be an adult in all dimensions. The society was required to learn multiple skills and be productive in all areas as there was no specialization.

Adeyinka (1993) summarized the purpose of indigenous knowledge as traditional education being channel that enable an individual to,

- Develop his/her potentialities to the full so that s/he could acquire knowledge and training in a profession and so earn a good living,
- Position himself /herself in the society into which s/he has been born to live,
- Cultivate good habits and develop the right attitude to life and work,

- Develop as a good citizen and to explore the world and find his/her own place in it.

According to Moumouni (1968) this type of knowledge is "the school of life" whose purposes were focused on producing an honest, skilled, respectable, person who will have rounded personality and developed his community as well as to improve its growth. Woolman (2001) added that every single knowledge and skill, learnt were used for protection, ensuring successful reproduction, shelter and acquisition of food. Woolman further stated that, other aims of indigenous knowledge were to build unity and harmony in society, to preserve the cultural heritage of their community and preserve its margins, to instil feelings of communal living and to prepare the young children for adult roles and status in future. Traditionally, a person is considered human only in the midst of others and seeks both individual and joint harmony as the main task in the process of becoming a true person (Sarpong, 2002). African knowledge is collective and community oriented in terms of its method of acquisition. In other words, it has a collective, practical and social slant.

The Content of Indigenous Knowledge

African knowledge does not divide the syllabus of its contents into disciplines such as, arithmetic, sciences, arts, agriculture and economics mentioning just few. This is contrary with formal education where its curriculum is divided into learning areas. According to Nsamenang (2005), indigenous knowledge is arranged to meet the youth daily routines, integrating skills, the livelihoods of their family and knowledge all aspects of life into a one syllabus. Nsamenang further stated that the indigenous contents is not documented but it is organized in order to fit the milestones of different developmental stages of a person that the culture recognizes (Nsamenang, 2005).

This traditional system of education is similar in most African countries. It involves attitudinal training, and intellectual in order to develop a person who is fully acceptable in the society. In addition, hunting, farming, cooking, carpentry and different kinds of games, including wrestling, marriage counselling and critical thinking form part of the content of indigenous education content at different developmental stages of the life of the children (Adeyemi and Adeyinka, 2002). The learning method was through participation, observation and imitation what adults are doing.

No learning of the history of indigenous knowledge in Africa is comprehensive without knowledge of the indigenous people in Africa before the introduction of western education and Christianity (Woolman 2001). This means that indigenous knowledge used its own principles on which it was organised. Which are preparedness, perennialism, holisticism, functionalism, and communalism. Sifuna, (2006), like Omona (1998) contends with Mush (2002), that all of the above philosophical bases were used to equip boys and girls with appropriate skills to be responsible adults in their community.

It is very important to note that there is so much to learn from the indigenous knowledge systems of indigenous people. Now that there is illuminating light on what indigenous knowledge is, academics, planers, researchers and policy makers should pay more attention when wanting to understand and adopt this vital knowledge that is threatened by extinction. This, therefore, requires one to appraise the Western methodologies that have been used to study this knowledge.

Prospects for Multicultural Methodologies

Appraisal of Western Research Methodologies

The pivotal tenet of the appraisal of western methodologies is to amplify that research can never be objectively observed from an outsider's perspective rather it must be observed from inside through the direct experience of the people. Therefore, the role of the researcher is to understand, explain and demystify social realities through the eyes of different African indigenous participants (Cohen *et al.*, 2011, p. 19). As such, reality should rather be interpreted through the meanings that people give to their life world.

It is clear that most researchers still use Euro-Western research paradigms and frameworks in their research, even if they have to empirically justify African Indigenous Knowledge Systems. This practice creates not only a methodological challenge, but also challenges the golden thread of a scientific research process. It creates a methodological challenge because it interprets African indigenous knowledge not through the meanings that African indigenous people give to their life world.

For some time, African indigenous knowledge has been researched using Euro-Western-centred methodologies. From the onset, there was a great deal of resistance from indigenous peoples as most of these researchers used ethical considerations that are foreign to African local communities. As such, the African indigenous knowledge holders emphasise some cultural concepts that are sacrosanct and indispensable to them. Resultantly, if such are not adhered to, this results and/or has resulted in numerous African indigenous peoples not disclosing the information or giving half-truths thus, the researchers produce findings that are a mismatch from reality. This is so as the ethical strategies and methodologies adopted are processed from a purely Euro-Western perspective, without respecting the community of origin.

Limits of Western Methodology

There is deplorable (abusive) application of methodologies if research is not grounded in an African Indigenous Knowledge Systems framework or paradigm and is processed from a purely Western perspective, without respecting the community of origin. If relevant cultural practices are disregarded, African indigenous communities take on responsibilities and distort or frustrate what the researchers seek. On that backdrop, using Western-centred ethics on African indigenous knowledge research is not only dehumanising but portray them as 'pawns'. As such, to correct the wrong and introduce an adaptive system which upholds the values that play a central role in African indigenous communities, methodologies used to process the data need to be used to be grounded appropriately.

In the famous book of Ngugi Wa Thiong'o's (1986), the author indicated that military conquest and political dictatorship directly and indirectly imposed control of African indigenous peoples' lands. Resultantly, military conquest and political dictatorship also led to mental slavery and has persisted to do so even after independence. The latter is attributed to why Euro-Western researchers collected research data amongst them in a 'random and *ad hoc*' manner. Part of Ngugi Wa Thiong'o's book unearths 'research fatigue' on the part of the African indigenous peoples (the researched) and probes on the authoritativeness of such studies. The 'Euro-Western tradition or knowledge is grounded in positivism, the notion that research is an objective and value-free activity that can make sense of human and natural realities'. In that way, a research methodology is an essential element in research as it characterizes set of conventions on how knowledge is collected, analysed and codified.

However, since time immemorial, Euro-Western methodologies have always been used to study Africa phenomena in most places. The issue of using Euro-Western methodologies was not malicious as there were no known or documented African Indigenous Knowledge methodologies. It is worth noting that Euro-Western methodologies are themselves meticulously processed to respect and uphold human values and norms. Nevertheless, their standards fall short of satisfying African indigenous ethics. As such, a hybrid (mix) of methodologies is advocated for. By a hybrid (mix) methodology, the authors refers to the use of Euro-Western and African indigenous methodologies. One thorny issue which remains not conclusively answered. What does it mean to rejuvenate and protect African indigenous communities in a globalizing world of hybridization, borderlands, and multiculturalism? In answering the latter, Tamale (2011) notes succinctly three reasons for adopting a hybrid (mix) methodology:

- Firstly, Western theoretical perspectives define the underlying rationale and practice of the phenomenon.
- Secondly, if the Western concepts and theoretical frameworks are discarded, we would spend considerable resources reinventing the wheel, which is an unnecessary enterprise.
- Lastly, there is a lot of sense in using existing theoretical bases as a starting-point and then correcting or revising them in light of the contextual evidence collected in current studies.

Concurring with Tamale's submissions, it is important that the framework or methodology that is employed is grounded in the African Indigenous knowledge systems paradigm, because the community of research should be respected. Worth-noting, these two systems now co-exist thus need to be respectfully compared and contrasted without restraint. Writing about the alliances between indigenous communities, Tuhiwai (2011) notes that:

Sharing is a good thing to do, it is a very human quality. To be able to share, to have something worth sharing, gives dignity to the giver. To accept a gift and to reciprocate gives dignity to the receiver. To create something new through that process of sharing is to recreate the old, to reconnect relationships and to recreate our humanness.

From the above comment of Tuhiwai, it is clear that the concept of sharing provides a platform of reflection especially on protocol and research tools. The adoption of Euro-Western methodologies with the exclusion of Afro-centred methodologies in sharing platforms is alleged to be toxic. It is toxic because it has a tendency of persistently altering the identities that are known and misses the whole essence of the research purpose. As a basic principle, researchers (especially 'outsiders') who come to African indigenous communities should be vigilant on how to design projects so as to safeguard the dignity and equity of their participants. If this is not taken seriously, it has potential of unveiling hard questions about how a researcher's positionality has causal effects.

The theme below, therefore, discuss the significance of African-centred methodologies and its variations as 'Africa-centred, Afrocentric, Africentric, Afrocentricity, Africentricity, and African-worldview' methodologies in the collection of data among African indigenous communities. African cultural practices and ethics are socially constructed and socially represented; some view it as evil and oppressive against women who are usually victims while some find solace in it.

Prospect for Afrocentricity Methodology

One hidden but public information about the reality of life is that communities that bred us as Africans have direct and/or indirect means of remote-controlling the values of those that live in them; whether good or bad. This, therefore, denotes a conceptual framework shared by a community of scientists which provides them with a convenient model for examining problems and finding solutions in relation to that community. Moreover, this entails relationships to the human and non-human environments. Such shapes one's view of the universe, one's conception of time and of space (Kugara, 2017). However, it should be made clear from the onset that the 'world is too complex to be reduced to a set of observable laws and generalisability is a less important issue than understanding the real conditions behind the reality' (Gray, 2014). It is germane that in authenticating African indigenous peoples' life, history and culture from the perspective of African indigenous people, reality should rather be interpreted through the meanings that people give to their life world. It is also through observing and learning the cultures and belief systems of the indigenous people that knowledge is developed and theory is built.

The book appraises the significance of African indigenous culture and values in methodologies in the 21st century by rigorously evaluating the nature and conditions of things in a holistic development of the community and discarding unprogressive issues using the Constitutional muster. This stance is opted for as it advocates the evaluation of African values and norms from the African point of view in a bid to establish and sustain the unadulterated African status and identity.

Smith poses tough questions in her introduction to pave way for study designs:

"How 'good' researchers ask questions like? 'Whose research is it? Whose interests does it serve? Who benefits...?' And further, how 'good' researchers need to answer the questions of their participants; the Maori ask, "Is her spirit clear? Does he have a good heart? Can they fix up our generator?"

Some kinds of 'behaviour, actions and conduct' (culture) are approved while others are widely disapproved of because of different cultural subscriptions. To add, the disapproval of some values are strict as they regard them as sacred. According to Twumasi (2012), culture denotes the kind of behaviour that is exhibited and expected from each individual. With the latter definition of culture in mind and having established the place of values in a culture, the themes below narrows down cultural values to the African context:

African Social Values

Social values can just be viewed as those convictions and practices that are practiced by a specific indigenous society. The general public has a method for directing the convictions and practices that are performed either routinely by its members or performed at whatever point the occasion demands. Subsequently, most African indigenous communities have celebrations, recreations, sports, songs and dances that are peculiar to diverse communities. These exercises are done by the general African indigenous community since they are believed to be vital. Most of these social values, particularly in African culture, cannot actually be isolated from religious, moral, political values and so on. These social values are sponsored by African standard laws. They involve those conventional affairs that a people see as essential for their significant survival.

African Moral Values

African culture is implanted in solid moral considerations. It has an arrangement of different convictions and traditions which each individual should maintain in order to live long and to abstain from expediting curses on themselves and others. Infidelity, stealing and different types of improper conduct are strongly discouraged and at whatever point an alleged guilty party denies a charge brought against him or her, he or she would be taken to a traditional practitioner or made to make a vow for confirmation of guiltlessness. In other cases, they are brought before the African traditional leadership to be tried and sentenced if found guilty. In addition, African adages and astute maxims have a rich storehouse of insight. The adages caution against African wickedness and are "therefore a major source of African wisdom and a valuable part of African heritage" (Mbiti, 1977, p. 8). African culture has an ethical code that disallows doing mischief to a relative, a brother, an in-law, an outsider and a foreigner, aside from when such an individual is associated with immoral conduct; and if that is the situation, it is prudent to avoid such a person.

African Mystical Values

Religion in African societies 'seems to be the fulcrum around which every activity revolves'. Subsequently religious qualities are not played with. African traditional religion, wherever it is practiced, has 'some defining characteristics'. For example, it has the idea of a Supreme Being which is imperceptible and indigenous. It holds a 'belief in the existence of the human soul and the soul does not die with the body'. African customary religion likewise has the conviction that great and awful spirits do exist and that these spirits are what make correspondence with the Supreme

Being conceivable. Most importantly, it holds an ethical feeling of equity and truth and the learning of the presence of good and wickedness (Umoh 2005, p. 68). 'African religious values seem to permeate every facet of the life of the African and the African believes that anything can be imbued with spiritual significance'. The worship of different deities on various days proceeds to demonstrate that the African individuals hold their religious values in high regard. The traditional practitioners (diviners) 'help to streamline human behaviour in the society and people are afraid to commit offences because of the fear of being exposed by the diviners'.

African Governance System

The African culture certainly has political foundations with heads of such establishments regarded as important people. The most vital thing about the African traditional society is that the political hierarchy starts with the family. It was believed that 'disloyalty to a leader was disloyalty to 'God' and the position of leadership was either hereditary or by conquest'. Moreover, concerning political qualities, one can see that it is inseparably connected with religious, social, moral qualities, etc. It is the political value that a people hold which makes them accord respect to their political institutions and leaders. It is the political esteem that a people hold which makes them respect and honour their political institutions and leaders.

Having understood the above, it is now vital that the authors focus on understanding African indigenous ethics that could be key in developing methodologies that suit and respect African indigenous peoples. This will facilitate an appreciative way to indigenous communities who are not only taken as tools of exploitation but important subjects in research.

Decolonising Process for Curriculum Change

When Eurocentric methodology was used by anthropologists and some social scientist it was acceptable. With a move to decolonize education there was a great deal of resistance from indigenous peoples as most of these researchers used considerations that are foreign to local communities. One would then argue that the African indigenous holders emphasise some cultural concepts that are sacrosanct and indispensable to them. Resultantly, if such are not adhered to, this result and/or has resulted in numerous African indigenous peoples not disclosing the information or giving half-truths thus, the researchers producing findings that are a mismatch from reality hence an advocacy to decolonize western methodologies.

Decolonizing research is a process for conducting research with Indigenous communities that places Indigenous voices and epistemologies in the centre of the research process (Smith, 1999). It critically examines the underlying assumptions that inform the research and challenges the widely accepted belief that Western methods and ways of knowing are the only objective, true science. Holding Western beliefs and methods as “the” true science marginalizes indigenous methods and ways of knowing by condescending them as folklore or myth. In Smith’s view of decolonizing research, the researcher should centre indigenous values and follow Indigenous protocols. This does not mean researchers should reject all Western methods and theories, as they may be adapted if deemed appropriate and beneficial by the local community.

Some researchers have noted that decolonizing research is not as much about the method, but more about the spaces that can enable the research process – and that through this process, researchers’ identities also become redesigned or transmuted. Scholars writing in this field argue that researchers should use an indigenous lens in all phases of the project to examine the choice of theoretical frameworks and methodologies they use and how research findings can be translated into actions that promote social justice. Below is the proposed model of how African methodologies ought to follow. Worthy noting is that the model was designed following the Vhavenda ethnic group of South Africa thus, may not be pontificated on other groups with different norms and values.

Proposed Model for Development of African Methodology

According to (Chilisa, 2012), research methods that are associated to indigenous knowledge systems (IKS) are either used or underused in IK research regardless of calls for decolonised research methods. So, when researching about African indigenous people, the authors consider items such as paradigm, people (*Ubuntu*) ethics, measure to ensure trustworthiness and analysis.

Afrocentricity

The authors choose to use Afrocentricity as it complements to qualitative methodology in an African context because the theoretical paradigm underlying Afrocentricity is reliable with the Indigenous worldview. Afrocentricity is a paradigm that hold that the researcher should see, study interact with people, life reality from African perspective rather than western perspective. In other words, in African methodology, the researcher should be sensitive to the cultural norms and values of the researcher. They should also appreciate their feeling and to be careful as well as to be responsive. It examines the African reality from the perspective of the African and places the Indigenous practice at the core, recognises the Indigenous voice and confirms the centrality of cultural knowledge as the place to begin to create a lively African approach to research (Mkabela 2005).

This diagram below depicts a summary of what usually happens when a researcher is conducting a study in a Vhavenda community. The research approach cannot be detached from the western because when engaging with human beings in a natural setting, the research is qualitative. Though the approach, the Tshivenda design will still be called qualitative research design. All community researches are case studies and are alternatively called community engagements researches because community members are in one way or another participating as co-researchers. This is wherein, in some cases, authors come up with participatory action research or learning because participants are also learning from the conducted research.

This book recommends that Western methodology in African indigenous research be acknowledged and be used interchangeably with indigenous methodologies. Although advocating for decoloniality in methodology, the process should be given time for transformation. This will then allow vernacular terms to be inserted so as to blend the methodology and avoid reinventing the wheel.

This proposed model just stands as a first step towards putting an African rubric to follow when carrying out research within African indigenous communities. To properly oriented the reader on the essence of this proposed model, the African ethics are discussed below;

Indigenous African Ethical Codes

Traditional ethics refers to norms, precepts, principle and moral codes which regulate the conduct and actions from individuals in African societies (Udokang, 2014). The said ethics helps to maintain and ensure social order and stability in African traditional communities (ATC). Some of the African scholars refer to ethics as morals. Every indigenous community in Africa has their own customs that guided them on their tradition and lifestyle. Aiken (2004, p. v) defines a person's ethical sense as an evaluation of what is good and what is bad in absolute terms. Ethics involves approving good things that people do, and disapproving of the bad things that people do. In any society there are rules and regulations that elderly people expect the youth to live by. It is when these rules and regulations are not taken into consideration that those not following them are regarded as immoral (Ladzani 2014).

Stanford Encyclopaedia of philosophy (2010) states that the ethics of society is embedded in the ideas and beliefs about what is right and what is wrong, what is good or bad character. I t also depended on the conceptions of satisfaction in social relations, social principles and norms that guide the conduct of people in that particular community. In Africa the communities have their own institutions of learning where they believe the younger generations will learn more about traditional ethics and customs. Institutions like family and traditional schools were regarded as the sole agents of transferring traditional knowledge, values, norms, believes and principles. The following are some of the African traditional ethics that are common in different Vhavenda communities, p.

Figure 1.

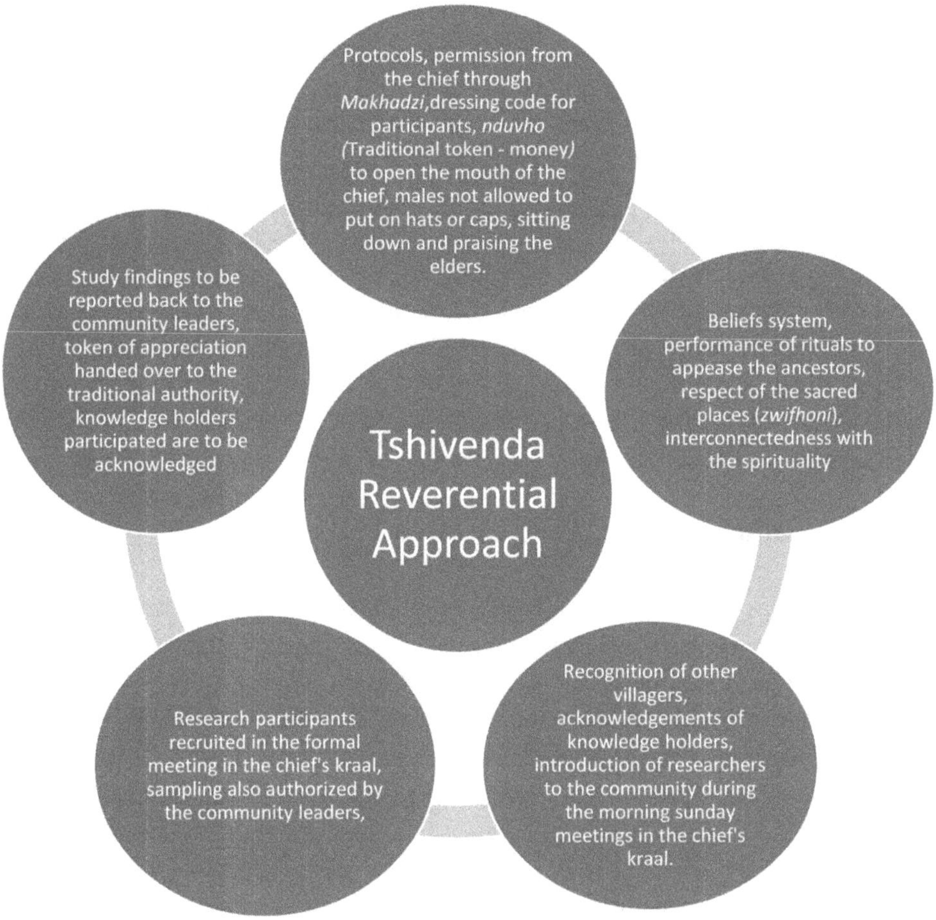

Ubuntu

Ubuntu is a lifestyle which is found amongst all Africans in their communities. The word Ubuntu is derived from a Nguni (isiZulu) aphorism, p. *Umuntu Ngumuntu Ngabantu*, which can be translated as "a person is a person because of or through others" (Moloketi, 2009, p.243; Tutu, 2004, p.25-26). Ubuntu can be described as the capacity in an African culture to express compassions. Africans believe that all humankind is one species, shares basic values, feelings, hopes and desires. Hospitality, sacred customs and generosity are some of the elements that defines Ubuntu.

An African child in every community is expected to qualify in all this elements. Of course, this is common in the Tshivenḓa/Vhavenḓa worldview which encourages deep respect for elderly people, who are referred to as those who brought us into the world (Khorommbi, 1996, p. 34). In research, humanity is one of the aspect that must be shown to the elders in order to get more information from them. Appreciating them in one way or the other to them means a lot. The above statement can be supported by stating that there are some instances wherein one should show respect and appreciation by bringing something tangible and of value to the elderly people that they are visiting. Hence, the Ubuntu philosophy is integrated into all aspects of day-to-day life throughout Africa and is a concept shared by all tribes in Southern, Central, West and East Africa amongst people of Bantu origin (Rwelamila, Talukhaba & Ngowi, 1999, p.338). In Tshivenda there is a saying which says "*phanda ha ndau a hu iwi u sina tshikuni*". Nobel Prize winner and former president of the Republic of South Africa, Nelson Mandela, describes Ubuntu as a philosophy constituting a universal truth, a way of life, which underpins an open society (Mandela, 2006, p.xxv).

Personal Behavior

Character in line with behaviour is identified as an individual's state of being, where the person's behaviour is governed by a ability to reason and think within the community context (Maphisa, 1994; Swarts & Davies, 1997, p. 290-296). Rational behaviour thus focuses on positive human values, such as love, sympathy, kindness and sharing. Respect refers to an objective and unbiased consideration of and regard for somebody's rights, values, beliefs and property (Eze, 2006; Tutu, 2004, p.26 & Yukl, 2002).

There is a western saying which says, character makes a person. This saying is also applied in African tradition, more especially, while arriving where the elders or other community members are. In case where a person is visiting the traditional leaders there are some protocols to be observed. The *makhadzi* or *vhavenda* is the one whom a researcher should talk upon arrival at the royal house.

Behavioural Patterns/ Actions

There are some sort of behaviour or actions to be shown while a person is approaching the elders in the community. If a male person is arriving in the vicinity of the elders he has to remove the hat if he is having it on his head, short trousers are also not allowed before the elders as a sign of respect. *U losha* (to bow or kneel down) is also a gesture to be practiced while greeting the elders. Males are required to squat before the elders with their hands together. In some instances where the researchers will be going to the chief he will be required to have a gift called *nduvho* . In a case of a female researcher she should not wear short or tight attire while in some traditional communities, even pants are not part of the female acceptable attire. The greetings gesture for females should also be well presented. The issue of gifts also applies to all people.

Venda people still believe in traditions which sustain them and gives them a good grounding in morality. That is why they still strongly believe in the concept of traditional leadership as the custodian of their morals. Their tradition or belief is that chiefs are installed by the ancestors and that makes them to a certain extent still obey the orders of the chief and respect him. The Venda people are characterized by a belief in group welfare and filial piety, and so the notion of traditional leadership is still central in their lives.

Lingua Franca

The issue of language plays a pivotal role in African research ethics. Most of the African researchers are ethnographic in nature and ethnography communication skills should be applied. Ethnography of communication is communication that consummates with the rules and values of a particular society. It was also referred to as ethnography of speaking by Hymes (1974). Ethnography of communication is an approach of understanding language use. How a person introduces himself or herself determines the degree of responses to the participants. As an investigator who bears the above mentioned qualities it is a must to put your matter logically with a very good native language of the participants. Every response that is given, whether right or wrong must be appreciated in a positive way.

Brown (1987) and Saville-Troike (1996) like many others explain communicative competence as a term in linguistics which refers to a language user's grammatical knowledge of syntax, morphology, phonology as well as social knowledge about how and when to use utterances appropriately. However, the above cannot be separated from ethnography communication. Communicative competence is dependent on the context in which the interaction takes place (Ladzani 2014).

Generally communicative competence is dependent on the context in which the interaction takes place. With regard to African traditional ethics in research, the researcher is expected to express himself or herself in a language well understood by the knowledge holder. The spoken language should also show respect in the side of the participant. Arriving on the research place the researcher must greet the people - one is expected to shout "Ndaa!!" (males) Or "Aa!!" (females) depending on gender. When they come to respond one must show a gesture of kneeling or squatting as a symbol of respect.

Belief Systems

According to Mashau (2007), the belief in the spirit world is a very integral part of the Vhavenda worldview. They believe in the world of the dead who, through their spirits, are in constant contact with the living. There are certain locations all over the Venda area that are known to be inhabited by the spirits. In actual fact, every chief has or had a forest or mountains in which the spirits of his ancestors are supposed to abide. Accordingly, many of these places are the actual burying places of the chiefs. West (1976, p.91; cf. Mashau, 2004, p.98) confirms this when saying, p. "Spirits of dead chiefs were believed to live in particular sacred groves; others in the mountains, in pools and in streams." Due to the above belief system researchers have to consider and abide with these cultural practices when doing their research in the tribal authorities.

Significance of Rite

There are certain rituals for communicating with the ancestors. If an individual is offended by his ancestors and he wants to communicate with them, the *makhadzi* can perform the ritual of spitting out water, (*u phasa madi*). He or she does so mentioning the name of the offending spirit, followed by the names of all ancestors and finish with the mention of the unknown, "... *na vhoiwe ri sa ni di*" for "... and you whom we do not know" (see Stayt, 1931, p.250; cf. West, 1976, p.91). Rituals were also conducted wherein one wishes to go in sacred places. This too ought to be respected by researchers – not saying some things, not wearing other attires, giving offerings, etc.

Sacred Places

As already noted, the spirits of dead chiefs were believed to live in particular sacred groves (zwifhoni); others in the mountains, in pools and in streams (see West, 1976, p.91). A typical example of the abovementioned is the Lake Fundudzi or Dzivha Fundudzi. Fundudzi is connected to the ancestor spirits of Netshiavha of the Tshiavha clan, who is the guardian of the Lake. It is also supposed to be inhabited by the spirits of other Vhatavhatsindi people (see Stayt, 1931, p.237; cf. West, 1976, p.91). The researchers should also consider the protocols when approaching these places for research purpose.

CONCLUSION

In conclusion, the chapter has endeavoured to examine the aspect of using a hybrid methodology in a decolonised African perspective, paying particular reference to living indigenous law. Furthermore, the chapter argued for the redesigning of ethical norms by including indigenous law, norms and values with an understanding of Africanism, departure from positivism and centralism which are products of Euro-Western views.

The chapter explored the current state of research within indigenous communities in South Africa. In doing so, the chapter highlighted that the research methodologies in South Africa find its roots and existence in Eurocentric myths, theories and ideologies. The chapter proceeded to explain the reasons why the current research methodologies ought to be decolonised. Therefore, it is recommended that that 'living' African indigenous law should be taught in research methodologies in order to equip students and researchers to comprehend indigenous law within the framework of indigenous communities. Upcoming lecturers of research and African indigenous knowledge holders ought to comprehend and conceptualise the methodology and development of indigenous law in a broad sense. This is essential to equip such future generations to have lens to observe indigenous law in its own right and not from the perspective of foreign systems. African ethical systems cannot survive unless it is scientifically taught in school in a proper way and context. The inclusion of living African indigenous law in the research curriculum confirms a move towards decolonising research methodologies through introducing a system of law that reflects, represents and relates to Africans.

The chapter recommends the adoption of an multicultural approach in research to decolonise Euro-Western research methodologies in South Africa. This will flow in interdisciplinary researches and other general researches. The National Science Foundation defined an interdisciplinary approach as:

"[A]n approach that integrates information, data, techniques, tools, perspectives, concepts, and/or theories from two or more disciplines or bodies of specialised knowledge to advance fundamental understanding or to solve problems whose solutions are beyond the scope of a single discipline."

The chapter argues that the adoption of an interdisciplinary hybrid approach offers a podium of understanding a phenomenon in all its manifestations. In that regard, there is a need to fuse African ethics and norms with other disciplines. It is the authors' view, that research is meant to address problems, however most of these problems lie beyond a single discipline and culture. This, therefore, confirms the need to integrate and understand life phenomena from different disciplines. Thus, the chapter advocates for an interdisciplinary hybrid approach that allows Eurocentric theories, views and ideologies to work together with Afrocentric theories, views and ideologies. African indigenous people's knowledge must not be subsumed or explained from the viewpoint of Eurocentric theories only. This kills the ability to search for the truth in scholarship. As such, data must be collected from all disciplines and cultures. This ensures checks and balances on the theories produced. The ordinary perception of an interdisciplinary hybrid approach relates to different fields, for example, sociology, anthropology and so on. However, this chapter aims to define an interdisciplinary hybrid approach from the context of fusing Eurocentric myths and theories with African theories and knowledge systems.

REFERENCES

Adeyemi, M. B., & Adeyinka, A. A. (2002). Some key issues in African traditional education. *McGill Journal of Education*.

Aiken, L. R. (2004). *Morality and ethics in theory and practice*. Charles Thomas.

Boateng, F. (1983). African traditional education, p. A method of disseminating cultural values. *Journal of Black Studies*, 13.

Broad-based Black Economic Empowerment Act 53 of 2003.

Brown, H. D. (1987). *Principles of language learning and teaching*. Prentice-Hall, Inc.

Cameroon, J., & Dodd, W. (1970). Society, Schools and Progress in Tanzania 1919-1970. London: Academic Press.

Chavunduka, M. (1995). The missing links. keynote address to the workshop on the study and promotion of Indigenous knowledge systems and sustainable natural resources management in Southern Africa. *Midmar KwaZulu-Nata, l*, 24.

Chilisa, B. (2012). *Indigenous research methodologies*. Sage.

Chinese Association of South Africa v The Minister of Labour (2008) ZAGPHC 174.

Dei, G. J. S. (2000). Rethinking the role of Indigenous knowledge in the academy. *International Journal of Inclusive Education*, *4*(2), 111–132. doi:10.1080/136031100284849

Fafunwa, B. (1974). *A History of Education in Nigeria*. Allen and Unwin.

Flavier, J. M. (1995). The regional program for the promotion of indigenous knowledge in Asia. In D. M. Warren, L. J. Slikkerveer, & D. Brokensha (Eds.), *The cultural dimension of development: Indigenous knowledge systems* (pp. 479–487). Intermediate Technology Publications. doi:10.3362/9781780444734.043

Hymes, D. (1974). *Foundations in Sociolinguistic: An Ethnographic Approach.* University of Pennsylvania.

Idang, G. E. (2009). The Mind-Body problem in African Culture. In *From Footmarks to Landmarks on African Philosophy*. Obaroh and Ogbinaka.

Khorommbi, N. L. (1996). *Echoes from beyond a pass between two mountains* (Unpublished Master of Theology dissertation). University of South Africa.

Knappert, J. (1995). *An encyclopedia of myth and legend: African mythology.* Diamond.

Ladzani, K. Y. (2014). *Moral regeneration in the lives of Vhavenda youth through indigenous knowledge systems: applied ethnography of communication-based approaches with special reference to Tshivenda.* Unpublished thesis.

Lincoln, Y.S., & Guba, E.G. (1995). Establishing trustworthiness. *Naturalistic Inquiry, 289*, 331.

Mashau, T. D. (2004). The Reformed Churches in South Africa's missionary pilgrimage among the Vhavenda people at Siloam. *Studia Historiae Ecclesiasticae, 30*(1), 90–107.

Mashau, T. D. (2007). Occultism in an African context, p. a case for the Vhavenda-speaking people of the Limpopo Province. *In die Skriflig, 41*(4), 637–653. doi:10.4102/ids.v41i4.324

Mashau, T. D. (2007). Occultism in an African context: A case for the Vhavendaspeak people of the Limpopo Province. *In die Skriflig, 41*(4), 637–654. doi:10.4102/ids.v41i4.324

Merriam, E. O. (2007). *Learning in Adulthood: A Comprehensive Guide.* Jossey- Bas.

Minnaar, A., Offringa, D., & Payze, C. (1992). *To live in fear: Witch burning and medicine murder in Venda.* HSRC.

Moumouni, A. (1968). *Education in Africa.* Praeger.

Mushi P. A K. (2009). *History of Education in Tanzania.* Dar-es-Salaam: Dar-esSalaam University.

National Science Foundation. (n.d.). *What is Interdisciplinary Research?* nsf.gov/od/oia/additional_resources/interdisciplinary_research/definition's

Nkondo, M. (2012). *Indigenous African knowledge systems in a polyepistemic world: The capabilities approach and the translatability of knowledge systems.* Paper presented at the Southern African Regional Colloquium on Indigenous African knowledge systems methodologies and epistemologies for research, teaching, learning and community engagement in higher education. Howard College Campus: University Of KwaZulu-Natal.

Nsamenang, A. B. (2005). Educational development and knowledge flow: Local and global forces in human development in Africa. *Higher Education Policy, 18*(3), 275–288. doi:10.1057/palgrave.hep.8300090

Omona, A.M. (1998). *Sociology of Education.* Department of Distance Education, Makerere: University Kampala.

Saville-Troike, M. (1989). *The ethnography of communication – An introduction.* Basil Blackwell.

Sefa, D. (2000). Rethinking the role of Indigenous knowledge in the academy. *International Journal of Inclusive Education, 4*(2).

Spitzberg, B. H., & Cupach, W. R. (1988). *Handbook of Interpersonal Competence Research.* Springer-Verlag.

Stayt, H. A. (1931). *The Bavenda.* Oxford University.

The Indigenous World. (2006). *The state has also established an Inter Departmental WorkingGroup on Khoe and San Issues.* Author.

Tutu, N. (1989). *The Words of Desmond Tutu.* Newmarket.

Uchendu, V. C. (1979). *Education and Politics in Tropical Africa.* St Martin's.

Warren, D. M. (1991). *Using Indigenous Knowledge in Agricultural Development.* Discussion Paper 127. Washington, DC: World Bank.

West, M. (1976). *Abantu, p. an introduction to the black people of South Africa.* Struik.

Woolman, D. C. (2001). Educational reconstruction and post-colonial curriculum development. A comparative study of four African countries. *International Education Journal,* 2(95).

KEY TERMS AND DEFINITIONS

African Indigenous Methodology: Methodology is the manner selected to obtain information for a study and describe the approach to the work. Thus, African indigenous knowledge is the African way of indigenous communities of obtaining information for study.

Decolonisation: Refers to a decolonisation process that indigenous populace embarked upon after attaining political independence with a view to transform structures and policies that colonised state used to subjugate indigenous people. In other words, decolonisation is however, not limited to political independence but rather it also extends to the way the previously colonised group was structured to think.

Indigenous: Occurring or living naturally in a specific area, such as native plants or animals (opposite to exotic); to be differentiated from 'endogenous', which means having its origin within a specific area (opposite of exogenous).

Indigenous Knowledge Systems (IKS): This term is used to explain the knowledge systems that are developed by a community as opposed to the scientific knowledge that is generally referred to as 'modern' knowledge.

Western Knowledge Systems (WKS): The WSKs is a term which comprises of the social norms, ethical values, traditional customs (such as beliefs) and specific artefacts and technologies as shared within the Western sphere of influence. In other words, WKSs refer to the content and context of knowledge systems driven by the values and cultures of Western civilisations.

Chapter 5
Constructivism, Pluralism, and Pedagogy From Below in India:
An Integrative Role of Educational Anthropology

Wahid Ahmad Dar
University of Kashmir, India

ABSTRACT

The chapter argues for a synthesis of principles of constructivism with the political and legal dimensions of educational pluralism. This synthesis has been argued to give a determining and central position to pedagogies from below in the process of framing curriculum and organizing teaching learning activities in Indian schools. Constructivism and educational pluralism represent two voices which assert for relying on psychological principles and countering the dangers of political hegemony in educational landscape, respectively. They both talk about the importance of pedagogy from below as a helpful reality focused toolkit for undoing the effects of historical social inequity and breaking the hegemony of dominant class in the process of knowledge construction and dissemination. Integrative role of educational anthropologists with the help of ethnographic research on native communities across the world in developing innovative methods on how can local community knowledge be integrated into formal curriculum and pedagogic practice has been highlighted.

DOI: 10.4018/978-1-7998-1249-4.ch005

INTRODUCTION

The practice of constructivist pedagogy has become inevitable keeping in view the theoretical and legal development around the world (Windschitl, 2002). Constructivism is a psychological theory, which unlike other psychological theories developed out of research on human subjects in real life settings than on animals in labs and in artificially simulated environments. Therefore, constructivism forms in real sense the psychological and anthropological principle of human learning (Ayers, 1991). Constructivists besides arguing for the active role of subjects, also highlight the importance of prior knowledge of students in the process of learning. Constructivism points towards creating an educational environment which forms a part of continuum from home and community to school culture (Windschitl, 2002). While constructivist psychologists argue for the role of community and local knowledge in school curriculum, pluralists assert that social reality is plural and school curriculum cannot be framed taking only the dominant culture into consideration. Therefore constructivism and educational pluralism represent two voices which talk of the need for relying on psychological principles and the dangers of political hegemony in educational landscape respectively. While constructivism emerged from psychological perspectives, pluralism emerged from legal and political perspectives on education. They both talk about the importance of pedagogy from below as a helpful reality focused toolkit for undoing the effects of historical social inequity and breaking the hegemony of dominant class in the process of knowledge construction and dissemination (Bernstein, 2000, 66-77).

The prime objective is to present a methodological analysis on how can the principles of constructivism and the legal and political perspectives of pluralism be integrated into a new framework called 'pedagogy from below'. Pedagogy from below shall be conceptualized as a methodological tool. This has been done particularly by drawing on worldwide research and knowledge of educational anthropologists. Educational anthropologists through ethnographic research on native communities across the world have developed innovative methods on how can local community knowledge be integrated into formal curriculum and pedagogic practice (see for example González, Neff, Amanti & Moll, 2006).

Critical Narratives from many south Asian countries indicates that importation of educational concepts and policy orientations have led to the diminished results of education, creating new forms of inequities and disadvantage. The super diverse nature of the region poses formidable challenges and opportunities for contextual innovation (Batra, 2019). This is very much important considering the challenges nation state like India is facing for developing culturally relevant pedagogy for the vast majority of culturally diverse communities and tribes. This will also be helpful for teachers and other educational practitioners who are facing practical and realistic difficulties in practicing constructivism and integrating local knowledge into formal instruction at the grass root level. This is evidenced by the fact that despite many policy documents and commissions recommending locality based learning, classroom instruction still remains in its age old un-psychological mode in India. As that is reflected by the poor educational performance of majority of the peripheral and marginal communities in India (Tilak, 2018; Thorat et al., 2017).

Perspectives From Below

Ommen (2001) provides us with a background for conceptualizing the 'perspective from below' as a methodological tool that should be employed to study hierarchical society like India. He refers to 'perspectives from below' as a necessary condition for providing the marginalized communities in India their legitimate share in the process of knowledge production (Oomen, 2001). His arguments are particularly for sociologists and anthropologist conducting research in Indian society, which he thinks is dominated by 'perspectives from above' (Kumar, 2016). Our concern here is how such legitimate representation in process of knowledge production is actually possible through the process of formal education. One of the dimensions of this representation can be various identity assertions from various linguistic minorities. Three language formulas which is in practice in India was a consequence of these assertions from various identity groups across India. But our concern here is a bit different and substantial. It is about promoting learning as a process of knowledge production (Paavola & Hakkarainen, 2005) by integrating the 'pedagogy from below' as means of formal instruction. Therefore in order to promote equality in the process of knowledge production and consequently equality to life chances 'pedagogy from below' must be considered as a vital tool. This would involve the contextualization of pedagogy at all levels of education, more particularly at the elementary level (Bernstein, 2000, 41-60). In the process of pedagogising knowledge, this would involve change in focus of reconceptualization that valuable knowledge is found only in a select group of experts or best practitioners toward the idea that useful knowledge is distributed throughout the whole of an organization (see for discussion, Bernstein, 2000, 41-61). Textbooks and policies designed at the national level, cannot be the only source of educating, rather students from different culture come to school with rich treasure of their family and community knowledge which can be a much effective source of educating (Dar & Najar, 2018).

The importance of pedagogy from below lies in the fact that it gives marginalized and peripheral communities their legitimate right not only in developing a sense of self-esteem and worth but to also actively engaging in the process of becoming responsible parts of a nation. Cutting out and silencing local perspectives and knowledge in debates about education does not suit democratic nation making (Dar &

Najar, 2018). While in other parts of the world social view of education has become an emergent discourse which emphasize for first hand understanding of the already existent indigenous cultural practices of literacy and learning, unlike what has been in vogue in the dominant policy perspectives of India (Street, 2002). The debates surrounding Indian education have been treating the internal diversity of the country's different people as a problem rather than an asset. Knowledge of 'other' peripheral communities can contribute not only to their own change and development but also of the nation as a whole (Mishra, 2019). Top-down approaches underrate the local agents and their knowledge and skills and pedagogies. As development pressures have taken prominence, such approaches have become focused on politics, which further tends to silence the pedagogies and practices from below without giving sufficient thought to learners and their social environment. State-centric approaches to teaching, learning and curriculum construction thus fail to consider what local support mechanisms may exist or need to be developed and what role individuals themselves can play in helping the state by becoming full-fledged persons at all levels, including responsible citizens (Hedge, 2018).

As already a kind of realization has evolved among educators and educationists by their encounter with reality of pluralism at the grass root level, that curriculum and pedagogy are highly dynamic and ever-changing entities and should be alive and organically connected to the immediate life of the child in and outside the school. This is because continued identification with ethos of one's own cultural group is not only necessary for the social supports but also basic to one's physical and emotional survival (Whitehead, 1978). In educational practice the shift represents the idea that knowledge base may not necessarily be restricted to teachers or prescribed curriculum but should be constructed and discovered taking the experiences of students and community members into consideration, what have been popularly called funds of knowledge (González, Neff, Amanti, & Moll, 2006; Moll, Amanti, Neff, & Gonzalez, 2005). This brand of educational practice is what we may call diversity responsive education, where due consideration is given to the knowledge from below as against the monistic versions of knowledge from above. In this context teacher's role is seen to nurture diversity and not to propagate a generalized dominant identity in the name of national curriculum (Sinha, 2016). The risks of Monist state-centric design of curriculum with an agenda of national systems of education have been sharply highlighted through a sustained critique by profounders of 'culturally relevant pedagogy' in anthropology and constructivist and 'socio-culturalism' scholars in psychology. Today more scholars turn towards recognition of pedagogic pluralism as an admittedly difficult but reality-focused tool kit for making teaching learning more aligned to the natural interests of the child.

Therefore In all its manifestations, socio-cultural contextualization cannot be undervalued for its potential in realizing the democratic and progressive ideals of education in a country like India, because of being in conformity with national ideals of promoting justice and equality. The ideals of a democratic nation can only be realized when such pluralistic version of educational practice are put into exercise. The basic feature of education in a democratic state is that, due consideration is given to views from below and all stake holders are part of the decision making process, also learning is based on the interests of the child (Dewey, 1970). Therefore emancipation of democracy in its true colors would only occur when pedagogies other than those which flow from above are given due consideration in process of educating nation. The under-representation of local knowledge systems in curriculum has been the key factor in limiting development of indigenous people, as educational success has been found closely connected to valuing minority or indigenous students cultural contexts and local knowledge in curriculum (Saigal, 2012). As These arguments are supported by the premises that Classroom teaching should be connected with the everyday understanding of children. It challenges the notion that local cultures are not rich in their knowledge content. It has been observed that Indigenous communities in India have an extensive and rich knowledge of mathematics and everyday science which does not find its place in the prescribed curriculum (Panda 2006). Same is the case with co-curricular and extra-curricular activities which must confirm to the tastes of children from native cultures. The silencing of pedagogies and epistemologies from below creates a gap between school culture and community culture of which leads to poor adjustment of children from disadvantaged sections in schools (Dar & Najr, 2017). The sense of separation between community life and school life also creates a sense of alienation among the children of marginalized communities, resulting into students experiencing discontinuities between their traditional way of life and the ways of the dominant culture imparted through schools and text-books (Kumar, 1983; Ramachandran & Naorem, 2013). Sometimes teacher's role as disseminating the dominant Indian value system through curriculum in the classroom causes students to dropout. These dropouts occurred because low-status students belonging to a lower caste and class background perceived a sense of discrimination with the more generalized kind of curriculum (Sinha, 2016). The generalized form of curriculum which is not based on indigenous knowledge systems may reflects colonial legacy in education (Ignas, 2004). Therefore one of the aspects of de-colonializing educational landscape would be to integrate local knowledge and contexts in curriculum & pedagogy (Ramani, 2011). A ridged national system of education besides reflecting a colonial legacy in curriculum designing, also puts forth the apprehensive mentality of policy planners towards contextualizing educational practice for its perceived role as fostering divisive elements within Indian nation state (Thapan, 2010).

Constructivism and Pluralism

The conceptual developments in the form of postmodernism, feminism, post colonialism, post structuralism, pluralism or critical race theories forward a common argument that the histories, experiences, cultures, and languages of marginalized students have been historically devalued, misinterpreted, or omitted within formal educational settings (McPhail, 2015). Some of them for example the critical race theories may demonstrate how the students of color are holders and creators of knowledge from below (Bernal, 2002). So the best thing postmodernism and many its allied concepts have in offer is the unique contribution towards the recognition of difference. As postmodernism may take many forms in art, literature, philosophy, social theory with everyone tossing it around like a beanbag, while aiming at different targets (Chin, 1989). In educational practice postmodernism is manifested as constructivism (Matthews, 2003). Constructivism is an educational theory which asserts that learner should be active in the process of learning and learning should start with the prior experiences of learners, may it be his home experiences or community knowledge. Therefore learning is a process of active construction of knowledge embedded in the culture of the learners and teachers are facilitators in the process. There are many versions of constructivism like Personal constructivism (Piaget, 1954), radical constructivism (Glasersfeld, 2013) and social constructionism (Vygotsky 1978) all share a common ground that students arrive in any learning situation with a range of prior knowledge and experience that influences how they respond to new information ((McPhail, 2015).

Plurality is reality and pluralism is an approach based on plurality. Plurality means the existence of multiple forms of cultures with their respective political, legal, educational and religious systems. Plurality and diversity are key to development and change in contemporary world (Thapa, 2019). Plurality has become a hot issue after the formation of nation states, as there are hardly any nation states without the existence of multiple cultures interacting in complex ways (Berry, 1974). Educational pluralism has been defined in different ways. School segregation is a form of educational pluralism in which different minorities are allowed to run their own schools in order to promote their own cultural identity. Pillarization as a version of educational pluralism existent in Netherlands (Sturm, Groenendijk, Kruithof & Rens, 1998) allowing educational autonomy to different subnational groups within the frame of national sovereignty, ensuring the integration of these groups to a minimal degree without jeopardizing the national Identity. Halstead (1995) terms this form of educational pluralism as voluntary apartheid, meaning that educational pluralism is inevitable for equitable treatment of all social groups in educational landscape. Educational Pluralism from legal perspective involves rights of minorities to promote their cultures through education, maintain educational institutions based on knowledge system of their cultures.

Educational policy planners in India have been apprehensive with the practice of educational pluralism as it has been inaccurately viewed as a source of intolerance and indoctrination. But empirical research says that a system of educational pluralism can be more efficient in fostering tolerance in a society than a uniform and state maintained system of education can. Newman, (1982) claims that pluralism is a necessary condition for tolerance and best way to develop the ability to accept pluralism is to institutionalize it. That may be why tension between local versus national curriculum have been a key feature of curricular reforms in India since independence. With the explicit emphasis on national system of education in Kothari Education Commission 1964-66, National Policy on education 1986 reforms initiated by the government of india have mostly emphasized the generalized, as opposed to the localized contextualization of curriculum development (Kumar, 1988; Tilak, 2018). Although, during fifties, curriculum policy in India was characterized by a pull toward local relevance under Gandhian basic education. Gandhi's plan for educational reform was defeated both by ideological opposition to his vision of a self-reliant rural society and by deliberate attempts to make implementation ineffective (Kumar, 1988).

The argument here is that educational pluralism is a logical necessity for constructivism to be practiced and vice versa. While education for monistic nation building starts with a national curriculum, educational pluralism argues for education to be locally grounded in community knowledge. Because in pluralism the assumption is that alternative perspectives are to be permitted into the discourse, displacing the dominant national or global hegemony of knowledge which is the domain of only a selected group of individuals (McPhail, 2015). Pluralists assert that knowledge is locally grounded and each culture has its own way of knowing the world, which is reflected in its unique way of making sense of the world or its distinguished world views. This way of knowing is conditioned by its unique historical development and socio cultural context (Ladson-Billings, 2000). In the most basic sense, pluralism is an acknowledgement of alternatives so that additional perspectives have the possibility of being understood (Chin, 1989). While pluralism in education has more relevance in legal and political contexts, like pluralism in education may mean legal provisions for representation for of minority cultures and languages in textbooks but constructivism is an educational theory which takes pluralism as its basic fundamental. Constructivism lays the psychological foundation for educational pluralism. Therefore with the arguments of constructivists educational pluralism doesn't merely remain a legal or political necessity but a psychological one as well. Although the pluralistic and constructivist forms of educational practice have been hailed by all corners is it social scientists and educationists or policy planners, it has been seen that in reality such ideas have not been put into practice, and the reasons are mostly political (Hedge, 2018). Practicing educational pluralism has been presumed to fostering sub national identities, many apprehensions emerge as such attempts are misconceived as danger to national and emotional integrity (Haydock, 2015). That is why the importance of education as an integrating force is seen as a critical justification for ignoring the value of difference, or the process of appreciating difference through education is undervalued to the celebration of universal (Haydock, 2015). Sometimes under the veil of national system of education hate is being generated between two rival nations like India and Pakistan (Lall, 2008). This is also why, while national curriculum framework recommends for a constructivist and pluralistic type of educational practice, the underlying assumptions behind such ideas are logically inconsistent with the basic idea of national curriculum (Tukdeo, 2019). For nationalist's educational practice in India is to actively protect and reinforce national identities and reform the local identities so as to coincide with those at the national level. In this scenario it is argued here that such a conception may have been true four to five decades ago but now the time has come when national pluralism is to be practically executed through educational institutes by practicing constructivism and consequently nourish the indigenous systems of knowledge. National pluralism pleads for pluralistic ways of accommodating multiple national identities within the same political community; it supports the view of allowing multiple identities grow with a multicultural nation (de Schutter, 2007).

Role of Educational Anthropology

The problem with practicing & recognizing 'pedagogy from below' at the grass-root level are both political and operational (Windschitl, 2002). On one side we have locality based pedagogy recommended by all major policy documents, while at the grass-root level you still have the undemocratic and un-psychological pedagogy in domination. The critical question that remains is why is there huge gulf between what is desired and what is practiced? The main reason is that proper channels have not been spelled out as to how 'pedagogy from below' can be integrated in formal teaching practice (Saigal, 2012). The practical side has been dealt by anthropologist in the form of developing culturally relevant pedagogy which aims at making teaching equitable for all (Ladson-Billings, 1995). These anthropologists take arguments from scholars who have been talking about the knowledge systems that emerge from a social, cultural, and political history different from the dominant culture, marginalized on the grounds of race, gender, culture etc. (Gordon, 1990). Anthropologist and educationist of this brand impress upon discovering and integrating knowledge and skills found in local households and communities that are more innovative and effective than the traditional forms of rote instruction (Moll et al.,

1992). Many anthropologists have engaged themselves in the process of how school experiences can be made more relevant to the local students like Au and Jordon (1981) observed that by permitting students to use talk-story, a language interaction style common among Native Hawaiian children, teachers were able to help students achieve at higher than predicted levels on standardized reading tests. This may be termed as instances of culturally responsive pedagogy, which refers to a dynamic or synergistic relationship between home/community culture and school culture of students (Ladson-Billings, 1995). These are called 'Funds of knowledge' which may be placed at the intersection of culture & psychology; these are historically accumulated and culturally-developed bodies of knowledge and skills essential for household or individual functioning and well-being. These funds of knowledge are the result of people's lived experiences, including their social interaction, religion, language etc. (Gonza´ lez & Moll, 2002; Moll & Cammarota, 2010) and are essential for a person's self-definition, self-expression, and self-understanding (Esteban-Guitart, Moll, 2014). In other words, funds of knowledge research are driven by an equity agenda that capitalizes on building on the students' and their families' knowledge and experiences as resources for schooling (ibid). However, the challenge lies in connecting these bodies of educational resources with teaching practice in order to connect the curriculum with students' lives. These 'funds of knowledge' represents practical, authentic and reliable channels to connect community knowledge to school curriculum. This also satisfies those who argue for alternative epistemologies in pedagogy because this besides socially and culturally contextualizing knowledge promotes nourishment of diversity. Besides contextualizing the processes of pedagogy and curriculum transaction (Vygotsky, 1978; Bruner, 1986; Lave, 1988; Sobel, 2004) and promoting learning as knowledge creation (Paavola & Hakkarainen, 2005), the funds of knowledge concept can be placed at the intersection of four broader objectives dealt in this paper these are: a) promoting critical pedagogies b) practicing educational pluralism C) practicing constructivism in education. It is logically consistent with and represents in its various manifestations all these theoretical connotations. Therefore in different ways educational anthropologists through ethnographic research on 'other communities' develop the channel and methods for practicing locality based learning and teaching (Barkin, 2003).

Role of Teacher

Teacher education programs represent one of important efforts committed to social justice and equity. Thus, focus has become the preparation of prospective teachers in ways that support equitable and just educational experiences for all students (Raina, 1999). The democratic polity of a country cannot be achieved through textbooks alone. The proactive engagement of the school teacher with processes of curriculum redesign is a necessary condition for its success (Batra, 2005). Most Indian teachers still teach in a traditional teacher-led way. Nationwide, there still remains a big gap between the constructivist ideas expressed in the National Curriculum framework and actual practice of curriculum transaction and teaching. The National Curriculum is far ahead of school reality (Haydock, 2015). Teachers are acting as passive elements representing the dominant values of society, contrary to this teacher have to play an active role in bringing about social change in terms of facilitating critical consciousness among students (Brinkmann, 2015). Teachers are educational leaders and their sociocultural experience and consciousness may not correspond to the student's consciousness but the context in which teaching and learning occurs provides ample potential for a critical environment to develop where a common consensus can be built by actively engaging both the parties in the process of knowledge creation (Sinha, 2016). Putting constructivist or metaphor of participatory learning to praxis in India would demand greater autonomy for teachers. Autonomy for teachers would imply greater professional self-reliance, demand for higher status, and local control (Dar & Jan, 2017). It would also mean conceding to the teacher the right to autonomy in matters pertaining to the choice of materials for teaching and in the construction of the daily curriculum (Kumar, 1988). A number of key structural adjustments would be needed to make the school pedagogy and curriculum locally relevant, particularly in the professionalization of teacher education system, devolution of powers to school management in matters of curriculum transaction, decentralization of examination system and infrastructure boosting of schools at all levels. The major challenge would be to get rid of the unhealthy over dependence of teachers and students on the dominant text book culture. This would need a decolonization of mind from the mental schemas that knowledge always flows from top to down and only a handful of experts at large can have the best knowledge needed for education of all. The prevalence of dominant text book culture reflects historical circumstances under which India's present education system developed, colonization affects the perception of knowledge in a society and ultimately the curriculum of its schools. Colonized people are forced to disown their image of the world, and to accept the image that the colonizer has. The colonized man is made to look at himself from the point of view of the colonizer (Kumar, 1986). Teacher education programmes will need to change the prevalent text book culture in India towards a one where school communities are centers of knowledge creation and teacher are personnel with the mandate of making what has been prescribed at the national level relevant at the local level. May it be discovering the funds of knowledge or making the curriculum and pedagogy locally relevant teachers have a pivotal role to play. The NCFTE 2009 and the draft NPE 2016 has explicitly recommended continuous and comprehensive evaluation system, which besides aiming at the reduction of examination load over students concedes a lot more autonomy to teachers in terms of student evaluation, which would eventually lead to teacher exercising their choice in deciding a locally relevant content of teaching within a limited frame of prescribed curriculum.

Having said this, we also need to caution ourselves of the risks of running radical form of constructivism. The radical constructivism by allowing teachers and students to construct their own knowledge, may result into disconnection from the subject matter, and from researchers in the fields of learning and instruction consequently inviting threats of prejudices, naive concepts, misconceptions, subjectivism, solipsism and uncommitted relativism (Terwel, 1999). Although the idea of the classroom as a community of inquiry is attractive, there needs to be a close connection to knowledge-producing communities (Seixas, 1993). Teacher's education programmes will be needed to train teachers in necessary research skills required to discover the local funds of knowledge. Proper balance between the global scientific progress and local discovery of knowledge needs to be maintained. Continuous professional development of teachers will be a necessary condition on which the ideas of critical pedagogy and constructivism may rest (Batra, 2014; 2005).

CONCLUSION

The essential principles of NCFTE 2009 reflects the vision of Educational policy makers and practitioners to realize the elusive but essential Goal of education for a democratic and pluralistic society. The principles that guide these perspective of the NCF 2005, NCFTE 2009 and draft of NPE 2016 such as connecting knowledge to life outside school, enriching the curriculum beyond the textbook, taking learning away from mere rote memorization and nurturing- identities within, visualize the progressive nature of education by wedding the philosophies of constructivism, pluralism, knowledge from below & critical pedagogy. This could be in the wake of challenge faced by many schools that serve culturally and linguistically diverse populations, in the form of disconnection between schools and students' home communities, which have both cognitive and affective implications for students (Bouillion & Gomez, 2001). Ideally, teachers are expected to make teaching relevant by identifying what the students know by matching their background knowledge with lessons. Teachers may find it difficult to bring the students into the lesson when the students do not share the middle class knowledge base informing the curriculum. (McIntyre, Rosebery, & Gonzâlez, 2001). NCFTE (2009) stress that for proper application of school knowledge to real settings, the connection of school life to community life is vital. NCFTE 2009 and NCF 2005 suggest inclusion of locally relevant content in the curriculum as well as pedagogy. All this is not possible without the intelligent agent of teacher. Teachers need to be equipped to select and organize subject content and learning experiences from the community for the classroom. Draft National policy on education (2016) recommends that the curricular underpinnings of NCF (2005) are valid and need to be implemented vigorously. Teachers and students should have access to multiple sources of knowledge rather than only the prescribed text book. This eventually will need developing the capacity of teachers in identifying entry points in the curriculum and textual materials which call for contextualization and development of appropriate teaching-learning sequences and episodes based on the identified local specifics (Batra, 2015). These specifics may include community knowledge about technology, local occupations both farm and non-farm, local folk culture including songs, festivals, fairs and games. As teachers develop curriculum materials and learning experiences, informed by the perspectives enunciated above they will also learn, through actual participation, the skills to identify and process the specifics for the purposes of meaningful curriculum transaction (NCFTE, 2009). India is a culturally plural society and the challenge of connecting the prescribed curriculum in books to children's immediate community knowledge and needs rests on the teachers. The proactive engagement of the school teacher with processes of curriculum redesign is a necessary condition for materializing of the vision of these policy documents. Therefore the exercise of curriculum renewal cannot afford to neglect the simultaneous need of transforming teacher education, if its purpose is to convert schools into sites of social transformation for a future egalitarian and just social order (Batra, 2005).

REFERENCES

Au, K., & Jordan, C. (1981). Teaching reading to Hawaiian children: Finding a culturally appropriate solution cited in Ladson-Billings, G. (1995). Toward a theory of culturally relevant pedagogy. *American Educational Research Journal, 32*(3), 465–491.

Barkin. (2003). Realist Constructivism. *International Studies Review, 5*(3), 325-342.

Batra, P. (2005). Voice and agency of teachers: Missing link in national curriculum framework 2005. *Economic and Political Weekly*, 4347–4356.

Batra, P. (2014). Problematising teacher education practice in India: Developing a research agenda. *Education as Change, 18*(sup1), S5-S18.

Batra, P. (2015). Curriculum in India. In Curriculum studies in India (pp. 35-63). Palgrave Macmillan.

Batra, P. (2019). Comparative Education in South Asia: Contribution, Contestation, and Possibilities. In Comparative and International Education: Survey of an Infinite Field (International Perspectives on Education and Society, Vol. 36). Emerald Publishing Limited. doi:10.1108/S1479-367920190000036011

Bernal, D. D. (2002). Critical race theory, Latino critical theory, and critical raced-gendered epistemologies: Recognizing students of color as holders and creators of knowledge. *Qualitative Inquiry*, *8*(1), 105–126. doi:10.1177/107780040200800107

Bernstein, B. (2000). *Pedagogy, symbolic control, and identity: Theory, research, critique* (Vol. 5). Rowman & Littlefield.

Berry, J. W. (1974). Psychological aspects of cultural pluralism: unity and identity reconsidered. *Topics in Culture Learning, 2*, 17-22.

Bouillion, L. M., & Gomez, L. M. (2001). Connecting school and community with science learning: Real world problems and school–community partnerships as contextual scaffolds. *Journal of Research in Science Teaching*, *38*(8), 878–898. doi:10.1002/tea.1037

Brinkmann, S. (2015). Learner-centred education reforms in India: The missing piece of teachers' beliefs. *Policy Futures in Education*, *13*(3), 342–359. doi:10.1177/1478210315569038

Bruner, J. (1986). *Actual minds, possible worlds*. Harvard.

Chin, D. (1989). Interculturalism, postmodernism, pluralism. *Performing Arts Journal*, *11*(3), 163–175. doi:10.2307/3245434

Dar, W. A., & Jan, K. (2017). Placing teacher at the junction of unity and diversity in India. *International Journal of Research in Engineering. IT and Social Sciences*, *7*(1), 56–59.

Dar, W. A., & Najar, I. A. (2017). Problems of Teaching Tribal Children—A Study on Tribal Schools of Khanshaib. *Academic Social Research*, *3*(4), 43–49.

Dar, W. A., & Najar, I. A. (2018). Educational Anthropology, Tribal Education and Responsible Citizenship in India. *South Asia Research*, *38*(3), 327–346. doi:10.1177/0262728018800759

De Schutter, H. (2007). Nations beyond nationalism. *Inquiry*, *50*(4), 378–394. doi:10.1080/00201740701489468

Dewey, J. (1956). *The school and society and the child and the curriculum*. University of Chicago Press.

Dixon, N. M. (1999). The changing face of knowledge. *The Learning Organization*, *6*(5), 212–216. doi:10.1108/09696479910299929

Gonzalez, N., & Moll, L. C. (2002). Cruzando el puente: Building bridges to funds of knowledge. *Educational Policy*, *16*(4), 623–641. doi:10.1177/0895904802016004009

González, N., Neff, D., Amanti, C., & Moll, L. (2006). Funds of knowledge for teaching: Using a qualitative approach to connect homes and classrooms. In *Funds of knowledge* (pp. 83–100). Routledge. doi:10.4324/9781410613462

Gordon, B. M. (1990). The Necessity of African-American Epistemology for Educational Theory and Practice1. *Journal of Education*, *172*(3), 88–106. doi:10.1177/002205749017200307

Gruenewald, D. A. (2003). The best of both worlds: A critical pedagogy of place. *Educational Researcher*, *32*(4), 3–12. doi:10.3102/0013189X032004003

Haydock, K. (2015). Stated and unstated aims of NCERT social science textbooks. *Economic and Political Weekly*, *50*(17), 109–119.

Hegde, R. (2018). Curriculum under Siege: Yesterday Once More? *Contemporary Education Dialogue*, *15*(2), 217–226. doi:10.1177/0973184918781218

Ignas, V. (2004). Opening doors to the future: Applying local knowledge in curriculum development. *Canadian Journal of Native Education*, *28*(1/2), 49.

Kuhn, T. (1970). *The structure of scientific revolutions* (2nd ed.). University of Chicago Press.

Kumar, K. (1983). Educational experience of scheduled castes and tribes. *Economic and Political Weekly*, 1566–1572.

Kumar, K. (1986). Textbooks and educational culture. *Economic and Political Weekly*, 1309–1311.

Kumar, K. (1988). Origins of India's" textbook Culture. *Comparative Education Review*, *32*(4), 452–464. doi:10.1086/446796

Kumar, V. (2016). How Egalitarian Is Indian Sociology? *Economic and Political Weekly*, *51*(25), 33–39.

Ladson-Billings, G. (1995). Toward a theory of culturally relevant pedagogy. *American Educational Research Journal*, *32*(3), 465–491. doi:10.3102/00028312032003465

Ladson-Billings, G. (2000). Racialized discourses and ethnic epistemologies. In N. K. Denzin & Y. S. Lincoln (Eds.), *Handbook of qualitative research* (2nd ed., pp. 257–277). Sage.

Lall, M. (2008). Educate to hate: The use of education in the creation of antagonistic national identities in India and Pakistan. *Compare: A Journal of Comparative Education, 38*(1), 103–119. doi:10.1080/03057920701467834

Lave, J. (1988). *The culture of acquisition and the practice of understanding*. IRL Report 88-0008, Institute for Research on Learning.

McIntyre, E., Rosebery, A. S., & González, N. (Eds.). (2001). *Classroom diversity: Connecting curriculum to students' lives* (Vol. 88). Heinemann.

McLaren, P. L., & Giroux, H. A. (1990). Critical pedagogy and rural education: A challenge from Poland. *Peabody Journal of Education, 67*(4), 154–165. doi:10.1080/01619569009538705

McPhail, G. (2016). The fault lines of recontextualisation: The limits of constructivism in education. *British Educational Research Journal, 42*(2), 294–313. doi:10.1002/berj.3199

Mishra, R. C. (2019). Cultural and Psychological Dimensions of Development of the Adivasi Communities. In J. Pandey, R. Kumar, & K. Thapa (Eds.), *Psychological Perspectives on Diversity and Social Development*. Springer. doi:10.1007/978-981-13-3341-5_7

Moll, L., Amanti, C., Neff, D., & Gonzalez, N. (2005). Funds of knowledge for teaching: Using a qualitative approach to connect homes and classrooms. *Funds of knowledge: Theorizing practices in households, communities, and classrooms*, 71-87.

Newman, J. (1982). *Foundations of religious tolerance*. University of Toronto. doi:10.3138/9781487577759

Oommen, T. K. (2001). *Understanding Indian Society: The Relevance of Perspective from Below*. Department of Sociology, University of Pune.

Paavola, S., Lipponen, L., & Hakkarainen, K. (2004). Models of innovative knowledge communities and three metaphors of learning. *Review of Educational Research, 74*(4), 557–576. doi:10.3102/00346543074004557

Raina, V. K. (1999). Indigenizing teacher education in developing countries: The Indian context. *Prospects, 29*(1), 5–25. doi:10.1007/BF02736822

Ramachandran, V., & Naorem, T. (2013). What it means to be a Dalit or tribal child in our schools: A synthesis of a six-state qualitative study. *Economic and Political Weekly*, 43–52.

Ramani, S. (2011). Decolonising Knowledge Systems. *Economic and Political Weekly*, 17–19.

Saigal, A. (2012). Demonstrating a situated learning approach for in-service teacher education in rural India: The Quality Education Programme in Rajasthan. *Teaching and Teacher Education*, *28*(7), 1009–1017. doi:10.1016/j.tate.2012.05.007

Seixas, P. (1993). The community of inquiry as a basis for knowledge and learning: The case of history. *American Educational Research Journal*, *30*(2), 305–324. doi:10.3102/00028312030002305

Sinha, C. (2016). Teaching as a political act: The role of critical pedagogical practices and curriculum. *Human Affairs*, *26*(3), 304–316. doi:10.1515/humaff-2016-0026

Sobel, D. (2004). Place-based education: Connecting classroom and community. *Nature and Listening*, *4*, 1–7.

Terwel, J. (1999). Constructivism and its implications for curriculum theory and practice. *Journal of Curriculum Studies*, *31*(2), 195–199. doi:10.1080/002202799183223

Thapa, K. (2019). Introduction: Psychological Perspectives on Diversity and Social Development. In J. Pandey, R. Kumar, & K. Thapa (Eds.), *Psychological Perspectives on Diversity and Social Development*. Springer. doi:10.1007/978-981-13-3341-5_1

Thapan, M. (2010). Imagining Citizenship: Being Muslim, Becoming Citizens in Ahmedabad. *Economic and Political Weekly*, 45–50.

Thorat, A., Vanneman, R., Desai, S., & Dubey, A. (2017). Escaping and falling into poverty in India today. *World Development*, *93*, 413–426. doi:10.1016/j.worlddev.2017.01.004 PMID:28966435

Tilak, J. B. (2018). Education poverty in India. In Education and development in India (pp. 87-162). Palgrave Macmillan. doi:10.1007/978-981-13-0250-3_3

Tilak, J. B. (2018). The Kothari commission and financing of education. In Education and Development in India (pp. 255-282). Palgrave Macmillan. doi:10.1007/978-981-13-0250-3_8

Tukdeo, S. (2019). Towards Reconstruction: Education in Postcolonial India. In *India Goes to School* (pp. 43–56). Springer. doi:10.1007/978-81-322-3957-4_3

Windschitl, M. (2002). Framing constructivism in practice as the negotiation of dilemmas: An analysis of the conceptual, pedagogical, cultural, and political challenges facing teachers. *Review of Educational Research*, *72*(2), 131–175. doi:10.3102/00346543072002131

KEY TERMS AND DEFINITIONS

Constructivism: In education refers to a learning theory which emphasises role of activity, prior knowledge, and culture in the process of teaching and learning.

Educational Anthropology: A sub discipline of socio-cultural anthropology which deals with education and studies role of culture in the process of education.

Pedagogy From Below: An emancipatory system of education where curriculum is more representative of the oppressed classes.

Pluralism: The belief in the existence and application of a wide variety of normative systems.

Politics of Education: The political implication of learning and education and vice versa.

Chapter 6
Indigenous African Knowledge System (IAKS) Ethos:
Prospects for a Post-Colonial Curriculum

Ntokozo Mthembu
https://orcid.org/0000-0003-4059-1215
University of South Africa, South Africa

ABSTRACT

This chapter discovers the limitations presented by narrow cultural and moral settings and the possibility of incorporating an indigenous African knowledge systems' (IAKS) ethos to redress past injustices, especially diverse cultural values experienced in countries in the 'global south'. However, the emergence of related protests in communities and student structures in education circles, such as calls to decolonize the curriculum and the #FeesMustFall movement. The effects of colonialism continue to be reflected in social structural settings that uphold those Aristotelian parameters that are notorious for marginalizing the knowledge of the 'other', specifically in the 'global south'.

INTRODUCTION

Worldwide the "postcolonial" era has brought possibilities of social change including in the education sphere, where curriculum change is high on the agenda, in particular in an African country like South Africa. However, the effects of colonialism continue to be reflected in social structural settings that tend to uphold the Aristotelian parameters, which are notorious for marginalising the knowledge of the "other" (Mthembu, 2018, p. 28). In other words, modern-day educational interpretations

DOI: 10.4018/978-1-7998-1249-4.ch006

and analyses of African people, phenomena and milieus continue to reflect a narrow cultural perspective that serves to marginalise indigenous knowledge systems and preserve social inequality among citizens (Do Vale, 2016, p. 600). This scenario tends to concur with the suggestion that colonial moral tenets do not adapt well to African-centred milieus, as they have altered cultural traditions in various parts of the continent in different ways; culminating in social crises on the environmental, political, social and economic fronts and undermining African wisdom and morals (Awajiusuk, 2015, p. 308).

To gain a better understanding of the underlying forces influencing the pace of transformation in the sphere of learning in this country, it is vital to consider how negative power dynamics have contributed to hampering transformation by failing to create an enabling environment where transformation can occur (Duthely, Nunn & Avella, 2018, p. 52). Currently, the absence of such a scenario is observable in the fact that, despite the adoption of various government policies, as well as interventions from stakeholders, to enhance the incorporation of indigenous African cultural, ethical concepts (*ubuntu*) in public administration structures; the transformation of the compensatory education system (including the curriculum) has not occurred. In reality, the incorporation of indigenous African knowledge systems (IAKS) in the education system is only happening on the fringes and many African scholars are still being denied the relevant support and acknowledgement (Mthembu, 2018, p. 194). This suggests that the compulsion to continue using foreign languages, such as English and Afrikaans, is the order of the day. However, for feasibility of this project, this chapter focuses on revealing the African-centred ethics and related guidelines that defines aspects of African research examination approaches. Thus, it is significant to understand the challenges and possibilities when it comes to the realisation of transformation goals in the governance structures in the post-colonial era and the adoption of relevant policies that will cater for a multicultural education system. In doing so, this chapter looks at the indigenous African knowledge systems' (IAKS) ethos, followed by strategies for a transformed curriculum, a transformative discourse and conclusion. The chapter serves to contribute to knowledge creation in this field of study, with a view to assist students, researchers, policy makers, stakeholders and IAKS specialists in articulating and understanding features and ethical standards of various knowledge systems especially African moral standards in a pluralistic education system.

BACKGROUND

For a better understanding of indigenous African knowledge systems, it is vital to define this type of knowledge. There are various definitions, with some focusing on related rites, relations, generational connectivity and linkages to other living creatures or fauna and flora. In other words, indigenous knowledge or indigenous African knowledge systems or African knowledge concepts are normally used interchangeably (Owusu-Ansah & Mji, 2013, p. 1). Since the discourse on indigenous knowledge systems has been ongoing in South Africa for more than 50 years, there has been varying definitions in this regard. Although they differ, all these definitions have the terms indigenous and knowledge in common (Mukuka, 2010, p. 3). Nonetheless, the debate on ancient Ethiopian Empire and historical spreading of ancient African people throughout the world cannot be exhausted, and it is worth mentioning that, although the debates rage on, the fact has been established that there were multiple dispersals predating 60,000 years ago in regions such as the Americas and southern and eastern Asia (Bae, Douka & Petraglia, 2017). It is therefore vital to be familiar with chronology in terms of time; although there are varying views in this regard, but for the purposes of this project, the chronological annotation that has been widely accepted in the academic circles is that African people revere the ancient natural order (or royal theocratic order) from their conception and their knowledge to uMdali uMkhulu-womkhulu (the Great Ancestor Creator Ptah or Ra/Raat - the energy that sustains life, an aspect of The Most High). Tehuti (the deity of African wisdom and intelligence); the scribe of ancestors was instructed, through Ptah's voice or word, to reveal the required moral standards through the art of holy writing *Mdw Ntru* (hieroglyphs) (*ibid*). Indigenous African knowledge is thus based on the aspirations of the ancient world order, the order of creation that respects the tidings of the Great Ancestor Creator, Ptah. Therefore, indigenous Africans have a common and traceable historical background and are referred to as the *amaNgoni* or keepers of the laws or the order of creation. Likewise, the Order of Creation or the Order of priest guides indigenous African people's way of life on Earth in dealing with social issues such as conception, human relations and the afterlife. Different knowledge's trace their origins from different locales, for instance, Western knowledge views Greece as its origin, while African knowledge reveres the ancient Kemet/Khemet/KMT, "the black land" [known today as Egypt], specifically Thebes/Wosè. This ancient city, also known as the university city, the city of a hundred gates, the eternal city and the citadel of the golden sceptre [with feathers on top], represents the unsurpassed ancient civilisation of the Ethiopian Empire (Nabudere, 2011, p. 86).

Furthermore, Nabudere (2011) argues that the process of conceptions of thought in Africa cannot be clearly professed without consideration of the role of the heart. Zulu (1993, p. 248) cited that Manetho, a Kemetic priest, noted in his writing, "Lost History of Egypt", that the Kemetic reign was characterised by 30 dynasties, in other words, Kemetic dynasties were subdivided. The pre-dynastic era, also known as Sebilian period, spans between 130 000–6000 BCE. According to the Pyrrhus, the Sebilian period is known for technological developments marked by the creation and use of small stone blades, the Stellar Calendar and the introduction of the first education system, in the ancient city of Elea, by King Akhenaton/Amenhotep IV (Ben-Jochannan & Clarke, 2017, p. 92). In other words, this era is linked to the discovery and the source of science, chemistry, mathematics, law and governance in the world. The next African periods are segmented into the Old Kingdom – First Intermediate, Middle Kingdom, Second Intermediate and the New Kingdom, also known as upper and lower Egypt in terms of locality. In addition, under the reign of King Menes (*circa* 3100 BC – 3038 BC), also known as Aha Mena and Narmer, these kingdoms were united to form a centralised government, *circa* 3200 BC, with Memphis as its capital (Ben-Jochannan & Clarke, 2017, p. 11).

Nevertheless, the era *circa* 2000 BC is considered the post-instability era, identified by the establishment of the Middle Kingdom (2134–1786 BC) and establishment of Wa-Set, also known as Wo-se' and Thebes (Zulu, 1993, p. 249). Nonetheless, the expulsion of foreign nomads, the Hykos, *circa* 1570 BC, led to the formation of the New Kingdom. This kingdom (*circa* 1570 – 1085 BC) was under the reign of Amenhotep I, II, IV and Amenhotep IV introduced monotheism. In addition, Tuthmosis I, II, III and IV, Makare Hatshepsut, Ramesses I and II (the Great), the construction of the Abu Simbel Temple and formal founding of Wa-Set/Wo-se'(Thebes) and Memphis as the intellectual, political, commercial and cultural centres of the ancient world (*ibid*). Subsequently, after the 20th dynasty (1200-1085 BC), Khemet was exposed to foreign invasion, spearheaded by Alexander, a former student of Aristotle's, in 332 BC, and the Arab/Muslim invasion, during 639-42 AD, which incorporated Khemet into the European and Islamic domains, respectively (*ibid*). In summary, the regions of the ancient Ethiopian Empire that span the whole ancient world, but, in this instance, the civilisations of various regions of Africa, are briefly sketched. Some of the regions that can be linked to the ancient civilisation of Ethiopia are as follows:

- Kerma Civilisation (Sudan & Egypt)
- Egypt Empire (North and South Kemet)
- Kingdom of Kush (Northern Africa)
- Kingdom of Axum (North-east Africa)
- Kingdom of Ife (Nigeria)
- Ghana Empire (Horn of Africa)
- Mali Empire (West Africa)
- Kingdom of Songhai (West Africa)
- Monomotapa Empire (Southern Africa)

The inseparable connection of knowledge to its people in a specific region makes it indigenous, because of its uniqueness to the cultural system of the specific community including Africans in general (Mukuka, 2010, p. 2). This was emphasised by Heeren (1832, p. 94) when he stated that:

Through which the Nile holds its course, above the confines of Egypt, with its monuments and inhabitants ... The southern frontiers of proper Egypt form merely a political boundary; the whole strip of land from a distant Meroe ... appears like a world by itself, cut off from the rest of the globe ... The same deities which were worshipped in Meroe, were worshipped down to this nethermost boundary... the same art in their buildings, their sculptures, and their paintings... just the same writing; just the same hieroglyphics ...

In this instance, the term "indigenous people" refers to the black African population that are scattered throughout Africa and in the diaspora during the Sebilian era and are called by different names, including the Grimaldi, Gqunukhwebe, Ngoni/Nguni, Ogoni, Lemba, Sotho, Hebrew, Tutsi, Tswana, Pedi, Hausa, Basotho, Khoisan, Xhosa, Jews, Swazi, Azanians, Ndebele, Zulu, Venda, Twa Tsonga and many more (Formicola & Holt, 2015, p.82). It has been established that the indigenous knowledge system embodies the cultural network and the lived experiences of indigenes, for instance, parents, in the pre-colonial African context, are the first tutors in the education systems and are responsible for imparting communal ethics, which is high on the agenda. In other words, African communities adhered to moral standards as prescribed in *Maat* (the pluses of truth, justice, righteousness, order, reciprocity, harmony and balance) by employing cardinal virtues in a monarchic communal system, which consist of the virtues of temperance, forthrightness, honesty, truth telling and integrity and valuing the sanctity of human life (Anyansi-Archibong, 2015, p. 252). Furthermore, IKS can assist in formulating pertinent developing alternatives, which will contribute towards the advancement of people-centred responsible development and sustainability (Mukuka, 2010, p. 4).

In order to better understand the issue under discussion, the inclusion of African-centred value systems, it is imperative to revisit the etymology of two fundamental terms, namely "ethics" and "morality". Kunhiyop (2008, p. 3) argues that these terms are interrelated, since both refer to an arrangement of ethical standards that guide the conduct of a person or people. As Kunhiyop (2008, p. 4) emphasises:

Morals and morality come from the Latin word, mos, meaning custom or usage, while ethics comes from the Greek word, ethos, whose meaning is roughly the same. So, it is hardly surprising that today, as earlier, these two words are often used interchangeably. When a distinction is made, "morals" nowadays refer to actual human conduct viewed with regard to right and wrong, good and evil, "ethics" refers to a theoretical overview of morality, a theory or system or code. In this sense, our morality is the concrete human reality that we live out from day to day, while ethics is an academic view gained by taking a step back and analyzing or theorizing about (any) morality.

Nevertheless, most literature tend to concur that an IAKS relates to a process, whereby a collective body of knowledge, practice and beliefs concerning the relationship of living creatures (including humans) with one another and with their milieu are considered (Gbolonyo, 2009). In addition, it is adaptive, as it has been transferred from past generations through cultural practices and forms part of an indigenous perspective or value system, which is embedded in ceremonial or religious practices. However, the role of the family in the socialisation process was paramount, since it played a meaningful role in tutorial activities and imparting community values to prepare young people to face challenges later in life.

In addition, the most significant features in the process of socialisation in Africa were age and gender, since these features determined the group an individual would be affiliated to (Momoti, 2002, p. 47). Socialisation in Africa followed a variety of ways, since it depended on the locality and rites of the specific community, but these rituals of socialisation were common to all the different cultural groups that had access to land (Carton & Morrell, 2012, p. 32). This is also emphasised by Mthembu (2017, p. 120), who cites Momoti (2002) and reports that:

From around six (6) years of age ... children are expected to join their parents in the chores which keep the homestead a well-functioning, orderly unit. Boys gleefully trot alongside the cattle, herding them towards their grazing and back into the cattle fold and learn the art of milking the cows.

Various empirical studies have linked different socialisation processes with childhood and adulthood (Richters & Waters, 1991, p. 185). The prevailing perspective infers that well-developed socialisation practices emphasise the sacrifice of personal interests for the benefits of the broader society (Richters & Waters, 1991, p. 195). This adverse approach is also revealed in the perceptive and emotional approaches stressed in the prevailing socialisation types. This approach, which is reinforced by guidelines and ethics, as well as the capabilities required to effectively engage, also play a meaningful role in the inherited ways of social culture and awareness; this is reinforced by discipline and rational initiation and standards (Richters & Waters, 1991, p. 196). Furthermore, there are two views in the current research on culture, the first view is based on the 19th century practices which stipulate that culture should be viewed as an independent "superphysical and superorganic" unit that accounts for its related rules, stages of improvement and inner self-motivations. Second view is based on the humanistic customs of the 18th century, which describe culture as the outcome of human exploration and innovation and thus as being subject to human regulation and commands (Mthembu, 2017, p. 38).

Perhaps, the relevance of the use of the notion of culture in this chapter emphasises the role of human intelligence and freedom and the activities of persons structuring and determining their own cultural destinies and the condition of their communities. Culture departs from the assumption that there is substantial merit in attempting to locate the natural conception and integrated logic of evolving traditions of learned, patterned, transmitted and widely shared human behaviour. In this chapter, culture is regarded as based on a transcultural, transtemporal and holistic context; and socialisation as a feature that concerns all cultures. It is noteworthy that today's young people experience fundamental changes in cultural transmission, to the extent that some of them happen to know more than their parents and elders, and can share this in the same manner that parents had traditionally inculcated their cultural values to young people (Ngaloshe, 2000, p. 55). For instance, respect is one element that was enforced in different ways in the community: the elders respected their king; likewise, the young people were expected to respect their elders, including their elder siblings, irrespective of gender; and a young person who disobeyed this rule faced the possibilities of being punished or regarded as an outcast (Mthembu, 2017, p. 38). Consequently, these differentiations resulted in various definitions of socialisation which suggest that: "We may define socialization as the process by which someone learns the ways of a given society or social group so that he can function within it" (Slee, 2002, p.10). Another definition, by Mthembu (2009, p. 7), states that the "socialization process is a variation relationship between a ... social and cultural environment, which prescribes the method and content of education, and the individual in whom experience is organised and internalized".

The concept of indigenous knowledge can also be applied in local decision-making processes (Mukuka, 2010, p. 2). The substance of indigenous knowledge is applicable to a broad range of subjects such as agriculture, animal husbandry, food preparation, local beliefs and rituals, faith and spirituality, education, institutional development, natural resources management, health care and many more (*ibid*). Indigenous knowledge has a spiritual and practical value to its practitioners (Owusu-Ansah & Mji, 2013, p. 2).

Therefore, transformation has provided an opportunity for various communities to discover and explore the possibilities of a transformed education system – one that includes the body of knowledge of previously marginalised groups. Different people from different social spheres (including in the domain of education) have diverse expectations for changes to the production or introduction of new knowledge, teaching and learning approaches (Nabudere, 2011, p. 23). Despite differing opinions, there is general agreement that the learning process moulds learners' perceptions of their milieu and of themselves by broadening their epistemology (Cassell, 1993, p. 96).

When looking at proposed practical changes to the governing moral codes, these involve following an inclusive approach aimed at changing the exclusionary politics of the traditional educational curriculum, which reinforces a one-size-fits-all agenda. This type of transformation agenda is mainly associated with the negation of a diverse scholastic environment and is deficient in laying the foundations that will encompass the curriculum objectives, that is, the use of relevant indigenous languages in the learning sphere; ethics; design; implementation and evaluation, including areas such as research (Anyansi-Archibong, 2015, p. 252). Thus, this chapter advocates the idea that the educational sphere in the postcolonial era should be transformed to be relevant to and reflect the aspirations of a liberated nation (Zuze, 2012, p. 45). Besides the lack of clear implementation in regard to transformation currently, details are lacking when it comes to IAKS-related ethical guidelines. Again, this scenario can be linked to the high student dropout rate, which makes the need for and implementation of an African-oriented education system more critical than ever before (Mthembu, 2018, p. 195).

MAIN FOCUS OF THIS CHAPTER

This chapter highlights the limits and the possibility of introducing indigenous African knowledge systems into curriculum, especially when effecting a pluralist "postcolonial" education system in a country characterised by a multicultural community. Curriculum development theorists state that schools operate as the propagators of specific values, not only through their curriculum content, but also through configuration procedures, cognitive patterns and various funding protocols and related policies (Pinar, 2009, p. 318). However, the increased demands for a pluralist or multicultural education system require a high level of understanding of and respect for varied cultural value systems, especially in regard to the previously marginalised communities (Cajetas-Saranza, 2015, p. 39). Nonetheless, such an aspiration becomes thwarted by the observably slow pace of transformation that tends to be characterised by two contending fronts: the integration of the previously marginalised knowledge systems and the demise of "authentic" academia. In fulfilling the aspiration of this chapter, Mthembu (2018, p. 30) argues that it is significant that the South African curriculum should be centred around the concept of *place,* because it is one of the fundamental aspects of curriculum development, specifically when describing a unique state of affairs. Thus, the pluralist approach in curricula development becomes relevant for a community with diverse cultural backgrounds like South Africa. In addition, major contemporary curriculum discourses encompass political, autobiographical, research methodological and ethical, phenomenological and gender features that are interconnected with a curriculum theory of place, as rooted in Habermasian's notion of social psychoanalysis and a literal conception of place. In other words, psychoanalysis functions as a determining element of a science that integrates self-reflection (Anyansi-Archibong, 2015, p. 251).

Strategies for a Transformed Curricula

When deliberating on curriculum change, the transformative learning perspective states that the process of "perspective transformation consists of three aspects: psychological (that focuses on the understanding of the self), conviction (that deals with belief systems) and behaviour (centred on understanding individual lifestyle)" (Heilman & Clarke, 2016, p. 42). The transformative learning perspective identifies four basic features that shape our approach to learning as part of the educational process (Mezirow, 2009; Heilman & Clarke, 2016, p. 42), namely individual experience, critical reflection, dialogue and action. Thus, the development of a realistic approach is significant to guarantee a sound relationship between learner and educator. In addition, the transformative learning perspective stipulates that people's perspectives change and thereby influences their new manner of understanding, knowing and perceiving (Heilman & Clarke, 2016, p.55).

Normally, research is understood as a vehicle that is used to understand a phenomenon and related analytical framework (Awajiusuk, 2015, p. 307). Awajiusuk, (2015, p. 307) argues that education is characterised by defined structures enabled by communication approaches that allow the individual to learn. In other words, the activities that take place in structured (a school) or unstructured (informal) institutions are known for imparting specific knowledge and skills. This suggests that educating an individual is a vigorous exercise with multidimensional aspects, commencing within the family, at home, and expanding to broader social spheres, where the emphasis is on social order, ethics, policy and place. Communal settings are characterised by aspirations to achieve the common good of community, while adhering to ancestral customs that facilitate a clear understanding of the "dos" and "don'ts" revealed through ethics and taboos.

According to Mezirow (1978) and Mezirow and associates (as cited in Heilman & Clarke, 2016, p. 55), transformative learning entails an individual's obligation to reflect on previously held beliefs, perceptions and assumptions. Furthermore, the transformative learning approach assists in identifying and understanding the issues that must be altered to achieve the desired outcome. Using participants' experiences to bring about anticipated curriculum change, the blended approach can be applied in three ways: first, by identifying intuitive experiences (seeing, listening and touching); second, by understanding emotional experiences (sadness, prejudice and a sense of accomplishment); and third, by exploring individual experiences (recollecting personal constraints and contrasting them with individual and community events) (Heilman & Clarke, 2016, p. 55).

Considering that the constraints in such a framework enable a broader personal understanding of the interspersed structures that influence curriculum change in a specific society, impediments to their success and any potential opportunities for development might be overlooked. Therefore, Freire (1979) and Beckmann (2016, p. 300) emphasise the adoption of a holistic alignment of teaching techniques aimed at encouraging engagement with other forms of knowing – both the effective and the relational - to promote transformative learning. The success of expressive ways of teaching in the learning session requires educators to work on their holistic consciousness. This will create a learning environment that takes the whole person into account by including rituals derived from the community, as they relate to each learner's experience, and implementing these through expressive activities such as storytelling and cooperative enquiry. This evokes knowledge that allows for greater interrogation, so that the learners become conscious of their feelings and their relationship with their surroundings through expressive representation (Mthembu, 2018, p. 30).

Basics for Indigenous African Knowledge Curricula

It is worth mentioning that indigenous African research paradigms encompass the decolonisation and indigenisation of objectives, which seek to bring the research process to the forefront (Chilisa & Tsheko, 2014, p. 222). When exploring the fundamentals of indigenous knowledge system theory, it is worth highlighting that it includes social capital, indigenous technology and that it is people-centred (Owusu-Ansah & Mji, 2013, p. 2). Social capital refers to the notion of a customary economy and involves substantial innovation in its endeavours to become accustomed to change, for example in the environment or in society (Mukuka, 2010, p. 4). Normally, the word technology is known to literally refer to the use of information or, sometimes, the use of equipment or management of the surroundings in various ways to create or develop, for instance, products, procedures, manufactured articles, cultural methodology and laws to equip a specific community to meet their livelihood needs in order to survive (*ibid*). In addition, it is also inventive and involves designing and using equipment, which ends up being incorporated in livelihood strategies.

Although technology has emerged from direct consultation with the natural environment, it also relates to biodiversity. In other words, technology consists of hardware - such as utensils, tools, instruments and energy sources - and software - that is composed of information, processes, ability and social structures - that put a focal point on a specific activity (Mukuka, 2010, p. 4). Therefore, indigenous technology could be implied to be characterised by the collective hardware and software of indigenous communities. It also forms part of the national heritage of the country and is a national resource that should be promoted, enhanced and utilised to the economic benefit of the nation (Owusu-Ansah & Mji, 2013, p. 2). The values related to indigenous African knowledge systems, such as ubuntu, are popular for their sustainable development, as it ensures that development that human centred and perpetuation of ancient indigenous African value systems and related social structures (Owusu-Ansah & Mji, 2013, p. 2).

Moral/Ethical Guidelines

African scholars, grounded in the virtues of the African value system, embody the following values in their daily lives including when conducting research in various social settings:

- **Ubuntu Principle:** The spirit of this African worldview includes wholeness, community and harmony rooted in cultural values. In other words, a person becomes human only when in the company with others and seeks individual and collective harmony as the primary responsibility in the process of becoming a real person. In addition, central to this African worldview, is the strong orientation to collective values and harmony rooted in a collective sense of responsibility – the "collective ethic" – which acknowledges that the future survival of the people depends on harmony through interdependence and interconnectedness (Owusu-Ansah & Mji, 2013, p. 2).
- **Monarchy System/ Dynastic Reign:** A political system that guaranteed equity and equality, which were demonstrated by the nature of the political system that was based on the notion of a dynasty that helped all African people to be of equal status.
- **Collectivism/Brotherhood/Oneness:** The ubuntu principle is re-emphasised by how people consider themselves when dealing with one another, which is expressed through language such as to address the other as "umfowethu" or "udadewethu", which means "my brother" or "my sister".
- Ungenzi komunye into ongafisi yenziwe kuwe (equivalent to: "Don't do unto others what you don't want done unto you.").
- **Wholeness Principle:** The restoration of health is sought through the biomedical approach. Wholeness is possible when affected relationships are mended and restored, because a person is healthy when she or he is in harmony with others. In other words, indigenous African knowledge and its method of acquisition, has practical, collective and social or interpersonal considerations. For example, the conceptions of intelligence, which is demonstrated in the practical, interpersonal and social domains of functioning,

which, in turn, varies from the cognitive, "academic" intelligence as conceived by Western concepts of the construct. Likewise, socialization process for the African young person is mainly peer-linked and participatory with minimal adult teaching (Owusu-Ansah & Mji, 2013:2). In summary, in light of the above moral standards, Zulu (1993, p. 254) submits that there are two sets of instruction that the African scholar must be grounded in and aspire to realise:

The learner or initiate should demonstrate an understanding of the seven liberal arts and ten virtues:

1. The liberal arts: grammar, rhetoric, logic, geometry, arithmetic, harmony/music and astronomy.
2. The ten virtues:
    a. Control of thought
    b. Control of action
    c. Commitment of purpose
    d. Identity with spiritual life to higher ideals
    e. Evidence of a purpose in life
    f. Evidence of a spiritual call to orders or the priesthood in the mysteries
    g. Freedom of resentment when under persecution or when in the wrong
    h. Confidence in the power of the higher consciousness
    i. Confidence in one's ability to learn
    j. Readiness or preparedness for initiation

Fundamentals of Indigenous African Inquiry Practise

It is argued that any paradigm that is based on indigenous knowledge principles has decolonisation intents, because of its motivating axiological (learning about the nature, types, criteria of values and assessment criteria, particularly in ethics) character. The Africanization process denotes a critique of a narrow cultural perspective in the education approach that tends to be characterised by Western language, thought and culture and academic intellectual imperialism. The Africanization process in research methodologies stipulates that scholars must incorporate scientific and academic analysis, that originate from their historical and cultural context and which forms part of the development of their paradigms. In addition, the Africanization agenda unleashes the indigenisation knowledge contained in languages, proverbs, folktales, stories, songs, music, taboos and artefacts, as well as the cultural and lived experiences, to realise new topics, themes, indigenous-centred conceptual frameworks, methods, processes and classifications of analysis that are not accessible through conventional methods.

Furthermore, the Africanization process enables the researcher to explore the ontological, epistemological and axiological views that guide the investigation process, based on an indigenous relational perspective. In other words, ontology focuses on enquiry about understanding personal existence in relation to the connections of human beings and connections with the non-living. In addition, it is normal that indigenes are associated with the land, earth, animals and other beings. Thus, the nature of "being" is summed up by the notion of *"ubuntu ngabantu",* which means, "I am, because we are", that incorporates humanness (Chilisa, Major & Khudu-Petersen, 2017, p. 327). Indigenous knowledge systems embrace methodologies that reinforce and promote social relevant enquiry, intended to contribute to problem-solving standards. Thus, the four "Rs" of relational accountability consist of: respectful representation, reciprocal appropriation and the rights and regulations framework.

A relational epistemology advocates for indigenous African ways of knowing that encompasses the manner and practices of conducting the investigation, networks, relationships, linkages and systems that form part of reality and its ways of being understood by its practitioners. Thus, Afrikology epistemology emanates from the African cosmology of connectedness and spirituality and it promotes harmony and balance, including critical inquiry of and openness to new ideas. Its assumption emanates from "we-ness" and "us-ness" (Chilisa1 & Tsheko, 2014, p. 329).

Afrikology Perspective

The Afrikology perspective pertains to a cultural integrative approach that involves the balance of complementary knowledge's from other cultural backgrounds and social justice elements and combines it with the best of indigenous methodologies with a decolonisation intent. A mixed style, geo-centric research approach is a balance of energies and golden principles of balanced harmony as an inclusive framework. In addition, in all relations including relationships with people of other cultures, the environment and the living and non-living. In other words, this philosophy encourages holistic, dynamic and sustained relationships for indigenous and non-indigenous scholars and communities to formulate a multi-epistemological enquiry partnership that advocates for equality of all knowledge (Chilisa & Tsheko, 2014, p. 223).

Afrocentric Method Termed "Afrocentricity"

This perspective complements qualitative methodology in social research in an African context, because the philosophical and theoretical paradigm underlying Afrocentricity is consistent with the African worldview. Therefore, Afrocentricity is a paradigm that has at its core the understanding of the African identity is centred on, located in and based on the African culture in all aspects such as spiritual, social, political and economic (Asante, 1987). Thus, the examination of the African reality, from the perspective of the African, tends to emphasise the African experience at the core; recognises the African voice and reaffirms the centrality of cultural experience as the place to begin to create a dynamic multicultural approach to research. In other words, Afrocentricity is not just about Africa or the African culture, but also about its fundamental purpose, that is, all cultural centres should be respected. In addition, to be "centred" means to be located as an *agent/agency*, based on equity and respect, to realise mutual exchange and synthesis of information. In addition, Afrocentricity, in research, promotes cultural and social inclusion, as opposed to scientific distance, and advocates for the use of tools and methods that are familiar to indigenous people who are being studied (Asante, 2007, p.15). Thus, indigenous people become active participants in expressing their views, wishes, concerns and they can pose research questions that they deem important to their cultural context and experience. Furthermore, social inclusion would enable the researcher and participant to be knowledgeable about the history, language, philosophy and myths of the people under investigation; and avoid misinterpretation, perpetuation of myths and researcher imposition. In summary, this suggests that the approach to data analysis is based on detail and is sensitive to the real social context and inquiries about the methodological foundations or chronological manner of ordinary social experiences. In other words, the investigator should be familiar with what he or she does and how he or she does it in a specific cultural context, the problem and related contextual dynamics.

Methodology

The research methodology should embrace African knowledge and methods of knowing that informs African investigation, so that it could be meaningful to its people and their context and will include, among others, the following:

Verbatim and Graphics

The indigenous knowledge system is typically oral and passed on from one generation to another in the context of community living and activities. Again, culture is viewed as the "window" through which a person observes, interprets and makes sense of his or her reality (Chilisa & Tsheko, 2014, p. 230). Thus, the inclusion of indigenous African knowledge in any investigation should concern the scrutiny of African reality from the perspective of African-centredness.

Ritual and Rites Custodian of Cultural Values

The complexities that are found in community ceremonies and rituals such as storytelling, proverbs, folktales, recitation, illustrations, sport, poetry, reasoning, riddles, praise, songs, word games, puzzles, tongue-twisters, dance, music and other learning-centred activities.

Critical Observation and Analysis

It is worth highlighting that African knowledge and methods have successfully guided people in all spheres of life including the spiritual, social, educational, agricultural, political and economic.

The Participatory Approach

This approach enables the researchers to learn from indigenous people's experiences and create a mutual working relationship, whereby indigenous people set their priorities and values in the research. Thus, individuals or communities that participate in the investigation are perceived as "participants" and equals, in both the process of research and the decision-making processes that are envisaged to contribute to policy change in their communities (Mthembu, 2011, p. 37). This research methodology emphasises the notion of "nothing about us without us", because of its collaborative nature that reinforces the African value of collective responsibility and centrality of indigenous African ideals and values as the frame of reference of conducting research from data collection to analysis and implementation.

Qualitative Method

The core to the Afrocentric paradigm is the inclusion or involvement of indigenous people in all aspects and stages of the research, from the beginning to the end, and the research project must be unique to the social structure and cultural values of African indigenous communities.

Spiral Methodology

A data collection approach involving the community, participants, researchers and decision-makers, all interacting in a synergistic and bidirectional manner with the emphasis on the African value of oneness.

Discursive/Conversational Method

In any community, each day starts and ends with a talking circle to gather the views of intellectuals concerning their daily experiences and to reveal their thoughts in conversations with their peers. The talking circle is used as a method to build individual trust and cohesion and to develop openness and confidence among people. In other words, talking circles are based on the premise of participants having respect for one another and serves as a data collection instrument similar to face-to-face and focus group interviews (Mthembu, 2011, p. 36). In African contexts, people normally form a circle when they congregate; that is, the fireplace, rituals and games. In other words, the talking circle symbolises oneness or egalitarian encouragement of sharing ideas; respect for one another's views, communality and a continuous and self-regeneration (Chilisa & Tsheko, 2014, p. 230). This approach is observable in everyday talking circles, storytelling and yarning, derived from indigenous customs, and are preferred to the typical interview methods, because they reflect the ideal of equality among the participants and emphasise the building of relationships and connectedness among people and of people with their milieu.

Indigenous Mixed Methods Research Approach

A mixed methods approach, offered in the indigenous research paradigm, integrates multiple ways of knowing and seeing the world, multiple points of view and multiple values. In other words, it promotes a multidirectional imparting and complementing of other knowledge systems between dominant and marginalised cultures. Thus, the basic requirement for effective implementation of the indigenous mixed methods approach requires the educator/researcher and the students/ participants to begin with a change of attitude, and to be critical of and challenge their own narrow cultural viewpoints and ways of knowing and perceiving the world of other communities (Mthembu, 2011, p. 37). Literature analysis does not only depend on written texts, but also on oral/unwritten texts of the people under investigation. Furthermore, a mixed methods approach emanates from a postcolonial indigenous paradigm, based on the everyday experiences of indigenous people, in conjunction with the theoretical perspective, derived from indigenous research and based on philosophical conventions about the nature of reality, knowledge, values and methodology (*ibid*).

The indigenous mixed methods approach is about inviting different voices to participate in a dialogue that embraces all cultures and promotes the social validity of research studies. From this perspective, there cannot be an indigenous research without including mixed methods research instruments. Both quantitative and qualitative approaches are indigenised. There are a variety of tools that can be used to build relationships between people, space, time and the environment and methods based on the indigenous epistemologies. Thus, the indigenous mixed methods approach requires that the data collected reach and draw from the past, occupy and inform the moment and project into the future (Mthembu, 2011, p. 38). In the research process, indigenous methods such as naming and storytelling are interlinked, and talking circles are central in this, since they are linked to the spiritual, historical, social and ideological aspects of the research occurrences. In other words, the mixed methods approach combines indigenous methods and qualitative methods to design a quantitative study instrument to measure the issues under investigation (*ibid*). It also combines quantitative data findings, derived from indigenous methods, and theory-based data, derived from the qualitative structured interviews, to develop a culturally relevant and appropriate intervention to promote social change (Chilisa & Tsheko, 2014, p. 228). Lastly, it also enables the identification of the beliefs and attitudes that influence the emergence of the situation that requires the intervention to alter the situation.

A Transformative Discourse: Reflections and Observations

A more encompassing approach is suggested to make changes to the South African curriculum as per the transformative learning perspective. Different types of educational resources are available and accessible to learners, but they depend on affordability as well as historical, linguistic and teaching techniques, all of which will determine the eventual level of educational transformation and extent of the benefits that previously disadvantaged communities can derive from them (Beckmann, 2016, p. 297). By removing the unstructured, yet significant, constraints, such as language, for South African learners, specifically those in previously marginalised segments of the population, it is envisaged that greater equity can be achieved, along with a decrease in the student dropout rate or an improvement in the pass rate.

In the South Africa, it is estimated that about 50% of black learners complete their secondary schooling, while 12% qualify to enrol at a university (Do Vale, 2016, p. 595). For example, in 2010, graduation rates were between 24% and 34% for black students and 37% for white students; postgraduate success rates were between 65% and 74% for black students and 80% for white students (Higher Education South Africa, 2014, p. 6). There is thus a need to consider aspects, such as moral cultural variation and variation in the teaching techniques, and to formulate new approaches to fund education-related activities when dealing with post-apartheid curriculum transformation.

The author posits that a mistaken conceptualisation of and errors in dealing with the differences between learners' cultural backgrounds (the Western vs the indigenous) calls for a transformed educational programme to attend to the injustices. Learners' prospect of academic success will be facilitated by offering them adequate opportunities to ensure equity and by bringing stability to institutions of higher learning. Higher Education in South Africa (2014, p. 5) contends that diversity is paramount for intellectual empowerment and new knowledge creation. However, the lack of diversity limits the scope of educational opportunities and the forging of acceptable social cohesion, which will enable institutions to achieve their objectives (Higher Education South Africa, 2014, p. 5).

Teaching techniques, such as the use of the mother tongue, will create a learning environment that recognises various cultural and moral standards. Also, the implementation of African centred values in the curricula will enable a more interactive relationship among learners from different cultural background and between education authorities/officials and government policymakers. Arguably, a curriculum based on multicultural education and related teaching techniques is needed for transformation. It will help to bring about inclusive education by recognising and solving ethical dilemmas in the learning sphere (Vittrup, 2016, p. 40). Some teaching techniques that focus on a specific moral aspect of research, analytical parameters and learners' abilities are discussed in the following section.

Challenges in Implementing a Multicultural Perspective

Contemporary structural changes in our society have created a need to consider multicultural viewpoints in the domain of education. Nonetheless, current educational settings are limited in that they do not encompass or acknowledge other cultural knowledge's. This puts the possibility of transformation at risk of being categorised as discriminatory and oppressive by individuals who still view their circumstances as disadvantageous (Mthembu, 2018, p. 30). However, the slow response to calls for a curriculum change and transformation towards a more varied cultural ethos, theory and practice (moving away from the present postcolonial education) remains a concern. Cajetas-Saranza (2015, p. 39) cites four historically constructed reasons that can be linked to the rationale behind stakeholders' failure to transform the education system, which remains incompatible with the curriculum change agenda and the aspirations of this country, namely, the stakeholders

1. Are satisfying the demands of proletarised individuals, people who are denied their way of life (self-reliance), but inculcating the idea that learners to depend on selling their labour power to earn a livelihood
2. Uphold neoliberal capitalist values
3. Discourage educators from inculcating local culture and using cultural studies as a form of pedagogy
4. Are dehumanising the self-discovery of indigenous people's own social and cultural value systems in accordance with the Aristotelian ethos

According to Garriott, Reiter and Brownfield (2016, p. 158), various studies concur that multicultural education is very important, notwithstanding the decline in modern expressions of racism and evidence of colour-blind racial belief systems in other quarters. A multicultural approach offers learners the opportunity to achieve equitable education and will bring about mutual respect and participation in a culturally varied society (Cajetas-Saranza, 2015, p. 39). There is a lack of available literature on the various ethical guidelines underpinning multicultural education, specifically in research into indigenous moral codes and related structural arrangements. There is also a lack of theoretical studies on the efficiency of interventions into learner outcomes based on indigenous knowledge systems (Vittrup, 2016, p. 37).

Garriott et al. (2016, p. 159) argue that there are unlimited numbers of studies related to the proficiency of shared approaches to multicultural education. Nevertheless, in testing three such approaches, Blincoe and Harris (as cited in Garriott et al., 2016, p. 159) discovered that participants revealed a noticeable weakening in their self-negation attitudes towards other racial groups. Garriott et al. (2016, p. 159) propose four levels of multicultural integration, concerning curriculum change:

- The *contributions approach*, which encompasses cultural texts or artefacts featuring in the curriculum.
- The *additive approach*, which involves adding content or themes to the existing curriculum.
- The *transformative approach*, which involves changing the structure and content of the curriculum to facilitate learners' exposure to concepts and issues from a cultural perspective.
- The *social action approach*, which focuses on inculcating, and endowing learners with, skills to effectively participate in social criticism and become agents for change.

Garriott et al. (2016) argue that the first two approaches are not effective in changing prevailing racial prejudices in the curriculum and may help to entrench stereotypes and delusions. Thus, it is vital that educators familiarise themselves with different approaches to dealing with multicultural issues. To deal with these inconsistencies effectively, policies and strategies must be formulated to satisfy different goals, which acknowledge certain limitations and challenges – specifically those related to improved opportunities, success, quality and diversity. Given the wide range of challenges and responsibilities, the methods of teaching and learning need to be aligned with contextual and environmental realities, conceptualisations and scholarship (Higher Education South Africa, 2014, p. 4).

Perhaps, in trying to understand the rationale behind the varying views, it is imperative to revisit the historic development of education, which is characterised by a curriculum that embraced a narrow cultural perspective (Kallaway, 2002, p. 2). Although, in this country, the Afrikaans- and English-speaking communities (of European descent) are in the minority, the education policy emphasises a curriculum that ensures the marginalisation of indigenous knowledge systems, labelling them as "unscientific" (Mthembu, 2018, p. 196). Over the past two decades, a multicultural education approach to curricula started to emerge (Garriott, Reiter & Brownfield, 2016, p. 158); however, a change in the curriculum approach – moving away from the theory and practice that characterised education under apartheid – remains to be realised. According to Connell (1993, p. 24), a conventional approach to the manner of doing research, teaching and assessment in the compensatory education system exacerbates the intellectual bankruptcy of disadvantaged nations.

For Beckmann (2016, p. 299) it is important to note that most viewpoints, which seem to be ill-defined and unsolicited, reflect certain attitudes, cultures and styles that obstruct the goal of effecting curriculum change inclusive of different cultural backgrounds. For curriculum change to be achieved, various issues need to be dealt with, including the transformation of institutional cultures and learning environments in the education system, as remnants of the many exclusionary, hurtful and anger-provoking elements of oppression of the past is still present today (Mthembu, 2018, p. 195). Thus, the aim here is to contribute towards redressing past injustices by identifying moral guidelines for an African-centred scholarship in areas such as student learning approaches and research standards.

To effectively deal with inequity and ethics, policies and strategies must be formulated to satisfy different goals, while acknowledging the limitations and challenges related to improving the curriculum. This will help to ensure student success, quality education and a diversity of viewpoints. A wide range of challenges and responsibilities, pedagogic methods and learning needs must be addressed to be in alignment with contextual realities, conceptualisations and scholarships (Higher Education South Africa, 2014, p. 4). Usually, education is viewed as a route to acquire the skills needed in a specific society – skills which will allow an individual to be able to perform particular tasks. However, the contemporary compensatory education system has limitations, which prevent individuals from becoming creative and self-reliant. The impact of colonisation on education has resulted in an alteration of social values, including a move away from lifelong education that enhances the fundamentals of higher consciousness to a materialistic lifestyle that promotes individual self-negation, specifically for indigenous African people, who have been forced into a labour-market-aligned value system (Mthembu, 2016, p. 4).

The nature of the education system is another factor that should be considered. In South Africa, the role of education, currently, is to serve labour market demands, which is widely acknowledged. Again, observations reveal that there are limitations to introducing relevant moral standards that encompass an indigenous knowledge-oriented education system. The nature of some of these social challenges relates to structural settings that encapsulate a mismatch between the skills endowment of black African individuals' self-understanding and their cognitive ability. The result is that Aristotelian principles continue to be upheld, especially when interpreting and analysing the totality of being African (Mthembu, 2018, p. 27). For example, the contradictory government policies related to language continue to categorise English as the official language for learning and conducting business, while encouraging the inclusion of indigenous knowledge systems. This situation suggests that, although individuals can achieve 'superior' skills and improve their employability, they remain alienated from themselves and are covertly exposed to a monophonic analytical framework.

Theoretically speaking, a multicultural education system can provide the type of skills required by the labour market, by developing the individual's ability to be self-knowledgeable, self-reliant and employable. Education should be accessible and should equip individuals with skills that focus not only on labour market demands, but also on their ability to be self-reliant. Of course, this recognises that there will be some variation between countries. Burns (2008, p. 11) emphasises that the current divisive social set-up has analogous connections to marginalisation – something which has discouraging effects, for the global South and African. Social experiences, therefore, often reinforce the aspirations that are embodied in the Western Enlightenment philosophy (an intellectual and philosophical movement that dominated the world of ideas in Europe during the mid-17th century and peaked in the 18th century) by employing a neoliberal agenda. Ownership of copyright, theoretical and analytical frameworks, domination of a single languages and moral standards that reinforce the domination of one culture's values (the European) over others (in this instance, indigenous African knowledge systems), saw Western-centred values reign supreme. For instance, the use of English for teaching and learning, which perpetuates the idea of one-size-fits-all and ignores the context and space of the individual (Mthembu, 2018, p. 27). In summary, issues of accessibility, the existing hierarchical social and economic structures and societal values have largely determined the acquisition of educational qualifications and societal values.

CONCLUSION

This chapter attempted to reveal some restrictions that inhibit the transformation of the curriculum and prospects for adopting certain approaches to rectify this situation in a changing South Africa. Challenges include the fact that there are etymological shortcomings in the process of realising aspired to changes in the curriculum. This cannot be divorced from the prevailing societal dynamics. The multicultural perspective represents the most appropriate approach to teaching and learning in the South African context: it will enhance learners' experiences and satisfy academic and societal aspirations. Some limitations relate to the challenges learners face in accessing education and the successful implementation of a transformed educational curriculum. This article argues that understanding the varied ethical codes for multicultural education is likely to become the cornerstone of education in the coming years.

The multicultural learning ethos underlying curriculum development depends on authentic learning, meaning inclusion and the usage of diverse languages including indigenous African lingua franca, ethics and value systems, together with physical, technical and practical skills. This chapter also tried to reveal some of the African-centred ethics and related guidelines that must be considered when conducting a study applying African value systems such as language in order to manifest a transformed education system. This article will contribute to development of guidelines and policies that will assist in the establishment of a transformed curriculum in the post-colonial era. In other words, a holistic approach is recommended that will attempt to deal with past injustices in terms of inclusion of the previously marginalised African cultural value systems in the education programme. Lastly, the success of the transformed curriculum, structure and relevant academic support for a diverse community, educators and learners should be inherent facets of the multicultural moral arrangements of any education system.

REFERENCES

Acharya, A. (2018). Introduction: Rethinking Agency and Change in Global Order. In *Constructing Global Order: Agency and Change in World Politics* (pp. 1–32). Cambridge University Press. doi:10.1017/9781316756768.002

Anyansi-Archibong, C. B. (2015). Challenges in Ethics Education: A Process and Content Analysis. In C. B. Anyansi-Archibong (Ed.), *Contemporary Issues Surrounding Ethical Research Methods and Practice* (pp. 250–273). IGI Global. doi:10.4018/978-1-4666-8562-8.ch011

Asante, M. K. (1987). *The Afrocentric idea*. Temple University.

Asante, M. K. (2007). *An Afrocentric Manifesto: Toward an African renaissance*. Polity.

Asante, M. K. (2015). *The History of Africa: The Quest for Eternal Harmony* (2nd ed.). Routledge.

Asante, M. K., & Abarry, A. S. (1996). *African Intellectual Heritage: A Book of Sources*. Temple University.

Ascione, G. (2016). Decolonizing the 'Global': The Coloniality of Method and the Problem of the Unit of Analysis. *Cultural Sociology, 10*(3), 317–334. doi:10.1177/1749975516644843

Awajiusuk, F. J. (2015). Prospects and Challenges of Teaching Religious Ethics. In C. B. Anyansi-Archibong (Ed.), *Contemporary Issues Surrounding Ethical Research Methods and Practice* (pp. 306–323). IGI Global. doi:10.4018/978-1-4666-8562-8.ch013

Bae C J., Douka K. & Petraglia M D. (2017). On the origin of modern humans: Asian perspectives. *Science, 358*(6368). DOI: doi:10.1126cience.aai9067

Beckmann, J. (2016). Exclusion and Education in South Africa: An Education Law Perspective of Emerging Alternative Understandings of Exclusion. *Education Provision to Every One: Comparing Perspectives from Around the World BCES Conference Books, 14*(1), 295–302.

Ben-Jochannan, Y., & Clarke, J. H. (2017). *New dimensions in African history: the London lectures of Dr. Yosef ben-Jochannan and Dr. John Henrik Clarke / edited with introduction by John Henrik Clarke*. Brawtley.

Burns, E. (2008). How Can a "Southern Theory" Perspective Contribute to New Zealand Counselling? *New Zealand Journal of Counselling, 28*(2), 10–24.

Cajetas-Saranza, R. (2015). Approaches to Integrating Multicultural Perspectives in PNU-Mindanao: Basis for Multicultural Education Program. *International Peer Reviewed Journal, 16*(1), 37–55. doi:10.7718/iamure.ije.v16i1.1122

Cassell, P. (1993). Problems of Action and Structure. A Giddens Reader. London: MacMillan. doi:10.1007/978-1-349-22890-4_3

Chilisa, B., Major, T. E., & Khudu-Petersen, K. (2017). Research Methodologies, a postcolonial, African-based relational paradigm. *Qualitative Research, 17*(3), 326–339. doi:10.1177/1468794117696176

Chilisa, B., & Tsheko, G. N. (2014). Mixed Methods in Indigenous Research: Building Relationships for Sustainable Intervention Outcomes. *Journal of Mixed Methods Research, 8*(3), 222–233. doi:10.1177/1558689814527878

Collins, A. K., & Millard, M. (2013). Transforming Education in South Africa: Comparative Perceptions of a South African Social Work Learning Experience. *Educational Review, 65*(1), 70–84. doi:10.1080/00131911.2011.648168

Connell, R. W. (1993). *Schools and Social Justice*. Temple.

Do Vale, H. F. (2016). Explaining Education Reforms and Decentralization in Brazil, South Africa, South Korea, and Spain. *Lex Localis: Journal of Local Self-Government*, *4*(3), 591–612. doi:10.4335/14.3.591-612(2016)

Duthely, M. L., Nunn, S. G., & Avella, J. T. (2019). Spirituality and Religion as Cultural Influences in Andragogy. In D. P. Peltz & A. C. Clemons (Eds.), *Multicultural Andragogy for Transformative Learning* (pp. 45–72). IGI Global. doi:10.4018/978-1-5225-3474-7.ch004

Formicola, V., & Holt, B. M. (2015). Tall guys and fat ladies: Grimaldi's Upper Paleolithic burials and figurines in an historical perspective. *Journal of Anthropological Sciences*, *93*, 71–88. doi:10.4436/jass.93007 PMID:25992636

Freire, P. (1979). *Pedagogy of the Heart*. Continuum.

Garriott, P. O., Reiter, S., & Brownfield, J. (2016). Testing the Efficacy of Brief Multicultural Education Interventions in White College Students. *Journal of Diversity in Higher Education*, *9*(2), 158–169. doi:10.1037/a0039547

Gbolonyo, J. S. K. (2009). *Indigenous Knowledge And Cultural Values In Ewe Musical Practice: Their Traditional Roles And Place In Modern Society* (PhD dissertation). University Of Pittsburgh.

Guilhermina, M. L. (2007). The Limits and Possibilities of ICT in Education. *Sísifo: Educational Sciences Journal*, *3*, 39–48.

Heeren, A. H. L. (1832). *Reflections on the politics, intercourse and trade of the ancient nations of Africa*. Oxford: D.A. Taalboys.

Heilman, D. C., & Clarke, D. L. (2016). Transformative Learning Theory: Perspectives on Nelson Mandela and Application for U.S. Learners in South Africa. In S. M. Tomlinson-Clarke & D. L. Clarke (Eds.), *Social Justice and Transformative Learning: Culture and Identity in the United States and South Africa* (pp. 40–57). Routledge.

Higher Education South Africa. (2014). *South African Higher Education in the 20th Year of Democracy: Context, Achievements and Key Challenges HESA presentation to the Portfolio Committee on Higher Education and Training Cape Town*. Retrieved October 17, 2016, from https://pmg.org.za/committeemeeting/17037/

Kallaway, P. (2002). The History of Education under Apartheid, 1948–1994: The Doors of Learning and Culture Shall be Opened. In P. Kallaway (Ed.), *History of Schools and Schooling* (pp. 224–242). Peter Lang.

Kunhiyop, S. W. (2008). *African Christian Ethics*. WordAlive.

Mezirow, J. (2009). Transformative Learning Theory. In J. Mezirow & E. W. Taylor (Eds.), *Transformative Learning in Practice: Insights from Community, Workplace, and Higher Education* (pp. 18–32). Jossey Bass.

Mthembu, N. (2009). *The State of Current African Borders and their Rationale: A Case of Recent Violence in Azania [South Africa] against Black African Populace*. Paper presented at the annual SASA Congress, University of the Witwatersrand, Johannesburg.

Mthembu, N. (2011). *Indigenous Knowledge creation Methodology*. Lambert.

Mthembu, N. (2016). Youth Challenges in the Transition to Adulthood in a South African Township. *Youth Voice Journal Online, 6*. Retrieved December 18, 2017, from https://youthvoicejournal.com/2016/12/

Mthembu, N. (2018). Ushering Transformative Change: Infusing De-colonized Cultural Changes into South African Curriculum and Andragogical Practices. In D. P. Peltz & A. C. Clemons (Eds.), *Multicultural Andragogy for Transformative Learning* (pp. 194–206). IGI Global.

Mthembu, N. C. (2017). *Black African Township Youth Survival Strategies in Post-apartheid South Africa: A Case Study of the KwaMashu Township within eThekwini Municipality* (PhD Thesis). University of South Africa, Tshwane.

Nabudere, D. W. (2011). *Afrikology, Philosophy and Wholeness: An Epistemology*. Africa Institute of South Africa.

Owusu-Ansah, F. E., & Mji, G. (2013). African indigenous knowledge and research. *African Journal of Disability, 2*(1), 1–5. doi:10.4102/ajod.v2i1.30 PMID:28729984

Parker, A., & Van Belle, J. (2017). The iGeneration as Students: Exploring the Relative Access, Use of, and Perceptions of IT in Higher Education. In Liebenberg, J. & Gruner, S. (Eds.), *ICT Education: 46th Annual Conference of the Southern African Computer Lecturers' Association*. Springer.

Pfeffer, T. (2012). *Virtualization of Universities: Digital Media and the Organization of Higher Education Institutions*. Springer-Verlag. doi:10.1007/978-1-4614-2065-1

Pinar, W. F. (2009). *Understanding Curriculum: An Introduction to the Study of Historical and Contemporary Curriculum Discourses*. Peter Lang.

Şimşita, Z. T., Günayb, N. S., & Vayvayc, Ö. (2014). Theory of Constraints: A Literature Review. *Procedia: Social and Behavioral Sciences*, *150*, 930–936. doi:10.1016/j.sbspro.2014.09.104

Slee, P. T. (2002). *Child, Adolescent and Family Development* (2nd ed.). Cambridge University. doi:10.1017/CBO9781139164948

Vittrup, B. (2016). Early Childhood Teachers' Approaches to Multicultural Education and Perceived Barriers to Disseminating Anti-Bias Messages. *Multicultural Education*, *37*(1), 54–67.

Zulu, I. M. (1993). *The Ancient 'emetic Roots of Library and Information Science. California Univ*. Afro-American Studies Center. Nov.

Zuze, T. L. (2012). The Challenge of Youth-To-Work Transitions: An International Perspective. In H. Perold, N. Cloete & J. Papier (Eds.), Shaping the Future of South Africa's Youth: Rethinking Post-School Education and Skills Training (pp. 42–59). Cape Town: Centre for Higher Education Transformation (CHET).

KEY TERMS AND DEFINITIONS

Africa: The continent that is known as the cradle of human kind whose majority of population consist of Black African people and the continent that boast with its historical inheritance as it is the first to experience civilisation prior to the emergence of western civilisation. In addition, it also experienced enslavement of Black African majority and intellectual imperialism that this article attempts to grapple with.

Coloniality: A concept that relates to the manner in which colonialism sustains itself despite the so called freedoms that were gained by the previously oppressed or colonised populace.

Culture: A set of values that are have been in practiced for a long time in a particular community and viewed as form part and parcel of identity of such community.

Education: A process that is used and accepted as a manner of inculcating specific values that govern that community. In other words, it is a set of social principles that individual is expected to emulate as a good citizen.

Ethics: Refers to a set of regulations or rules that are normally accepted as the guide in a particular community or structure to affect a particular behaviour or outcome.

Indigenous African Knowledge Systems (IAKS): Refers to a holistic approach when it comes to a variety of mechanisms that are used to create knowledge in the Africa sphere.

Morals: Denotes a particular standard that is perceived as acceptable in any society or must be emulated by all community members.

Socialization: A process whereby social values are imparted in particular young people to enable them to be in a better position to meet all challenges in the journal of life.

Section 3
Transformative Learning and Indigenous Knowledge Systems

Chapter 7
Towards an African-Centered Model of Learning:
Africanism

Rendani Tshifhumulo
University of Venda, South Africa

Faith F. Musvipwa
https://orcid.org/0000-0002-7691-9721
University of Venda, South Africa

Tshimangadzo Justice Makhanikhe
University of Venda, South Africa

Livhuwani Daphney Tshikukuvhe
University of Venda, South Africa

ABSTRACT

This chapter interrogates the current teaching and learning strategies in South African higher institutions of learning in relation to the negligence of cultural aspects. For many years now, the general population in South Africa agitated for the amendment or annulment of the current education system as a means of restoring the African norms and values. The suspension and/or expulsion of some African students from schools due to their adherence to African values is a vivid example of how the education system undermines African cultures and traditions. The chapter deliberated on story telling as one of the methods that is relevant for Indigenous Knowledge Systems. On the central argument of the chapter the authors conversed on colonialisation and its impact on education in South Africa. OBE and the revised national curriculum statement and the crises in South African education system are discussed. The chapter ends by interrogating the possibility of using African Languages in the school curriculum.
DOI: 10.4018/978-1-7998-1249-4.ch007

INTRODUCTION

This chapter interrogates the current teaching and learning strategies in South African higher institutions of learning in relation to the negligence of cultural aspects. For many years now, the general population in South Africa agitated for the amendment or annulment of the current education system as a means of restoring the African norms and values. The suspension and/or expulsion of some African students from schools due to their adherence to African values is a vivid example of how the education system undermines African cultures and traditions. Regrettably, with the issue of African cultures, they are taught as archaic, attributed to contradict the systematic, scientific and technological developments in a society. Resultantly, because these African cultures are not systematically written, it threatens this knowledge with extinction.

The results of such teaching and learning are expected to impact the African continent. However, it is deplorable and heart-breaking that students still use Western research paradigms and frameworks in their leaning, even if they should empirically justify African Indigenous Knowledge Systems. This system creates not merely a methodological challenge, but it justifies an abusive application of knowledge hence, challenges the golden thread of a scientific research process and purported learning outcomes. The first part of the chapter therefore, focusses on the African-Centred Theories to teaching and learning in relation to Indigenous knowledge systems. The second part of the chapter zooms into colonialism in Africa and how it affected African ways of knowing. In addition, the third part of the chapter looks specifically in Post-colonial and apartheid South Africa and its crises in education system. The last part of the chapter interrogate the importance of language, culture and development in teaching and learning processes.

The English word 'education' is often taken to refer to the formal systems of schooling originally introduced to Africa by colonial administrators, and further developed by post-independence governments. An examination of its original meaning, however, reveals something quite different. Senge (1990) highlights the fact that education is derived from two Latin words: educare, 'to rear or foster', and educere, 'to draw out or develop'. Education thus incorporates all the processes of raising up young people to adulthood and drawing out or developing their potential to contribute to society, that are traditionally found in rural communities.

This should not be interpreted as meaning that literacy, numeracy and the acquisition of new languages are unnecessary. No society can exist in isolation: people have always sought ways to communicate with one another and to trade in goods and services, and this has never been more important than it is today. In an increasingly interdependent world, it is as essential for us to be fluent in the languages of international economics and politics - in order to defend our rights and demand development on our own terms - as in the languages of animal tracks, bird calls and weather patterns.

What is currently missing, in most societies, is a system of teaching and learning that can combine the two. African children are either kept in their home environments, missing out on the 'modern' aspects of education, or (increasingly) forced into full-time formal schooling, missing out on the 'traditional'. The latter often furthers the neo-colonial mentality by building aspirations of urban life and encouraging young people to believe that they have no future in rural communities.

The Concept of Indigenous Knowledge Systems

Indigenous knowledge (IK) is defined as a unique cumulative body of knowledge generated and evolved over time and possessed by people belonging to a particular geographic area enabling them to benefit from their natural resources (Guchteneire, Krukkert & Liebenstein, 1999; Kaniki & Mphahlele, 2002; Sen, 2005). IK is unique to that particular geographic location and is passed down from generation to generation. IK can be differentiated from Western scientific or modern knowledge in that the latter is developed and generated by research institutions, universities and private businesses (Agrawal, 1995; Kaniki & Mphahlele, 2002). Indigenous Knowledge is developed outside the formal education system, it does not necessarily operate within formal organisations and is mainly transferred by word of mouth (Guchteneire, Krukkert & Liebenstein, 1999; Kaniki & Mphahlele, 2002). For the purpose of this paper, Grenier's (1998) definition of indigenous knowledge will be used: "the unique, traditional, local knowledge existing within and developed around the specific conditions of women and men indigenous to a particular geographic area".

The terms traditional knowledge (TK), indigenous knowledge (IK), and local knowledge generally refer to knowledge systems embedded in the cultural traditions of regional, indigenous, or local communities. Traditional knowledge includes types of knowledge about traditional technologies of subsistence (e.g. tools and techniques for hunting or agriculture), midwifery, ethnobotany and ecological knowledge, traditional medicine, celestial navigation, ethnoastronomy, the climate, and others. These kinds of knowledge, crucial for subsistence and survival, are generally based on accumulations of empirical observation and on interaction with the environment.

In many cases, traditional knowledge has been orally passed for generations from person to person. Some forms of traditional knowledge find expression in stories, legends, folklore, rituals, songs, and laws. Other forms of traditional knowledge are expressed through different means

Indigenous knowledge (IK) is broadly speaking, the knowledge used by local people to make a living in a particular environment. Terms used in the field of sustainable development to designate this concept include indigenous technical knowledge, traditional environmental knowledge, rural knowledge, local knowledge and farmer's or pastoralist's knowledge.

Indigenous Knowledge (IK) can also be broadly conceptualized as the knowledge that an indigenous (local) community accumulates over generations of living in a particular environment. This definition encompasses all forms of knowledge, technologies, know-how skills, practices and beliefs that enable the community to achieve stable livelihoods in their environment.

Wade Davis takes about a human 'ethnosphere', which is the sum total of human culture: all the thoughts, dreams, myths, inspirations, and intuitions that we have had since the dawn of consciousness. He argues that this is equally important to the world's well-being as biological life (which we know as the 'biosphere'). Whilst both face threats, however, Davis points out that not even the most pessimistic biologist would claim that half the biosphere is facing extinction – yet that is exactly the figure that applies when we are talking about the ethnosphere.

Afrocentricity

Although the abstract noun "Afrocentricity" dates back to the 1970s, it was popularised by Asante during the 1980s, when he developed epistemological and methodological foundations for an Afrocentric curriculum based on an African perspective but aiming at global understanding.

As an academic exercise, Afrocentricity is defined in terms of the methodology, theory, and ideology that should be employed to achieve its objectives towards attaining the proposed change. Methodologically, Afrocentricity is intended as an answer to the intellectual colonialism that undergirds and serves to validate political and economic colonialism. In regard to theory, it places African people at the centre of any analysis of African phenomena in terms of action and behaviour.

In terms of its aims and objectives, the Afrocentric movement comprises a series of activities by some concerned African and Afro-American scholars and educators directed towards achieving the particular end of ensuring that the African heritage and culture and its history and contribution to world civilisation and scholarship are reflected in the curricula on every level of academic instruction. Advocates of the Afrocentric ideas request a reconstruction and rewriting of the whole landscape of human history in its explanation of the origin of mankind, and the origin of philosophy, science, medicine, agriculture, and architecture.

Afrocentricity is only one of several cultural perspectives from which multiculturalism in education is derived, noting that Afrocentricity is not the opposite of Eurocentricity, nor does it seek to replace Eurocentricity. According to him, Afrocentricity is constructive primarily because it does not deny others their place. It is based on harmonious coexistence of an endless variety of cultures. Equally, there can be no true multiculturalism without Afrocentrism.20 Afrocentricity does not deny "the right of Europe to view the world from its cultural centre". The gist of the argument by Afrocentrists is that the European view must not be imposed as universal. Just as much as Europeans, Africans are entitled to give their own perspective on the African experience, making Afrocentricity an exercise in self-knowledge. Afrocentricity becomes valuable also for Europeans by giving them a new perspective, an ability to see from different angles and by so doing, putting them in a position to explore different views and bring new perspectives.21 Afrocentricity is not the opposite of Afrocentricity but a particular perspective for analysis. Consequently, it becomes necessary to examine all data from the standpoint of Africans as subjects and human agents, rather than as objects in a European frame of reference.

He brings forward the argument that Africans (at home and abroad) must look at knowledge acquisition, development and circulation from an African perspective. The theory suggests that people should look at higher education in terms of teaching and learning from an African viewpoint. He continues to argue that we misunderstand higher education in Africa when we use language, teaching and learning which are foreign. When Africans view themselves as centred and central in their own learning, they see themselves as agents of change, actors, and participants rather than as marginal and on the periphery of political or economic experience.

We cannot shy away from the reality that the view that anything associated with culture and hereditary values is pagan, and thus backward is still widely spread both globally and in the African continent. It is interesting to observe that even Africans themselves believe that learning on their own languages is disadvantageous. Advocate of this view argues that if a child learns for instance mathematics in Tshivenda that child cannot be competitive in the globalised world. Currently a lot of African parents spends a lot of cash in private schools for their children with a belief that they will know English or any former coloniser language better and they will be competitive enough in the globalised world. This is the kind of thinking that advocate of Afrocentricity is against. They argue that as Africans we should be able to uphold and be proud of who we are. Language is one of the most significant part of our identity as Africans which we should strive to develop.

The key indicator of a culture's health is its language. As Dr Renée Grounds, the lead language instructor and youngest fluent speaker of the Euchee tribe, says: Our heritage languages, our original languages are probably the most critical markers of the health of our communities in terms of our cultural well-being. And that relates to understanding who we are, that relates to understanding what our place is in the world, how we are to conduct ourselves; the special things that have been given to us by the Creator is all carried in the language. Judged along these lines, traditional societies are under serious threat. Roughly one language dies out every 3 months, which means that by the end of this century, if nothing is done to halt the decline, over 3000 languages will have died out. In the short term, it is estimated that of the world's current languages, around 450 are spoken by fewer than 10 people. These are the ones on the front line of extinction.

This chapter argues that if a person learns a particular subject in their language, that person get to understand the subject matter better as opposed to when the person learn using a foreign language. In Africa and in South Africa there is a general belief that mathematics and physical science are the most difficult subjects in schools. We agree with this view on the basis of understanding that before a learner struggle with the content being learnt he/she struggles with a language in which the subject is taught. It is for this reason that we have very few medical doctors, chartered accountants, engineers of different kinds and in all the professions that requires mathematics and physical science to gain access to them. Countries that have a surplus of the skills and professions mentioned above are known to have the kind of education system that prioritise their own languages as languages of medium instruction within their institution of higher learning. For instance, countries such as China prioritise mandarin as a medium of instructions in their schools and they learn English just as a subject or area of study whereas in most African countries particularly those that were colonised by England uses English as medium of instruction. In addition, if you go Spain, they use Spanish, Denmark they use Danish and in Germany they use German language in their institution of higher learning. We are of the view that if African countries want to develop faster, they should prioritise their own languages in their education systems to an extent that they facilitate learning using their own language in all levels of education and that any foreign language of choice can be made optional and not compulsory.

Storytelling as a Tool to Facilitate Learning

History also teaches us that Africans are known to have learned everything they knew using different methods of teaching and learning and in all the methods storytelling was the most used method. According to Craig (1996) storytelling is a process where a person (the teller), using vocalization, narrative structure, and mental imagery, communicates with the audience who also use mental imagery and, in turn, communicate back to the teller primarily through body language and facial expression in an ongoing communication cycle. This implies that the storyteller must at all times monitor the reaction and non-verbal responses from the audience so as to see how they are getting the story and also find a way for them to participate either through singing or other means. Storytelling is co-creative and interactive. It is one of the most powerful forms of art/communication known to humans and this explains why it possesses such great potential as a teaching-learning tool. Storytelling is aimed at helping leaners grow into adults who participate actively and competently in the development process with their communities, country, and continent and as global citizens.

Africans are people are globally known for their storytelling because they used it to pass knowledge and information from one generation to the next. It is a time of the day which is meant for sharing ideas and this has a lot of benefits to the people. It makes them coexist peacefully. At the dawn of the night, children gather beside their peers and the elders to hear amazing stories. In this, they also educate them about life, marriage, and their culture. It is a way of education which plays a vital role in the moulding of minds of the people. This is the major time where peers share ideas and the elderly share what they know with their people.

Storytelling is one of the oldest methods of teaching and learning in African traditional societies. This was supported by Baker (1977) argued that storytelling is older than history and is not bound by one civilization, one continent, or one race (Baker/Green, 1977). Pellowski, (1990) found that storytelling was practiced for religious and secular reasons as early as 200 BC. References are given highlighting the use of storytelling by such philosophers as Plato and Aristotle between 400 BC and 300 BC as a means of educating and socializing the child.

It should be noted that storytelling is an interesting way of conveying information and it allows learners to effectively reflect on it and the knowledge from that will be remains in their mind and heart. This view is supported by Boud et.al (1985) as cited by McDrury, & Alterio (2001) that defined reflection as 'a generic term for those intellectual and affective activities in which individuals engage to explore their experiences in order to lead new understanding and appreciations. African indigenous communities have been using this method to teach and learn various aspects of their culture, norms and skills for their own survival. It is for this reason that this chapter suggest the use of storytelling in classrooms within higher education environment. The chapter also suggest this method because some educators, find it extremely difficult to convey the information in an interesting way that can reflect to their students learning. Again, because learners and lecturers are emotionally involved with and often enjoy storytelling, it can help learners develop a positive attitude towards the learning process. Storytelling is known to produce a sense of joy in language and words that is usually missing in the classroom. It implies that a story should be told in a language that the audience are familiar with and well understand for the story to be effective in terms of facilitating learning for leaners. This chapter is mainly to discuss about the infusion and respect of cultural practices that can improve teaching and learning within institutions of higher learning in Africa and South Africa. It is also to rediscover the usage and importance of storytelling as an integral part of African culture in order to improve the technique of teaching and learning as well as to provide effective learning to the student. We know zoom into education during the apartheid regime to see how it influenced education system in South Africa.

Afrikaner Nationalist Education and Apartheid

Afrikaner education can be traced from the 1600s when it was minimal. By the 1800s it was still in the hands of parents and the church. After the Anglo-Boer War between the British and the Afrikaners, the defeated Afrikaners became disgruntled. The Boers/Afrikaners were opposed to the British system of education because they saw it as a way of alienating them from their own cultural practices. The Afrikaners then decided to establish their own schools based on Christian National Education (CNE). Article 15 of the CNE policy of 1948 explains the basis of apartheid education (Msila, 2007:148-149). We believe that the calling and task of White South Africa with regard to the native is to Christianize him and help him on culturally, and that this calling and task has already found its nearer focusing in the principles of trusteeship, no equality and segregation. We believe besides that any system of teaching and education of natives must be based on the same principle. In accordance with these principles we believe that the teaching and education of the native must be grounded in the life and worldview of the Whites most especially those of the Boer nation as senior White trustee of the native. The above objectives of education established by Afrikaner nationalists are very similar to those of the missionaries: the politicization of education and the abuse of religion play a role in both instances (Msila, 2007:149).

Apartheid education was a practice of maintaining that status quo and of preserving the master-servant relationship between the Africans and the whites. It was intended to "entrench apartheid capitalism", as was noted at the conference on People's Education in December 1985. Peter Kallaway (1988) points out that apartheid education not only domesticated the people but indoctrinated them as well. In the white schools in South Africa, the government policy in the past included various "educational programmes" where the young students would be indoctrinated in what was referred to as "veld schools" (literally translated as "bush schools"). This was done on the pretext of nature study. In the veld schools, the white boys were taught how important it was to protect their land from other people such as the blacks and communists, and these schools also enhanced the superiority complex of the white South African over the African.

Bantu Education for black South Africans had been a means of restricting the development of the learner by distorting school knowledge to ensure control over the intellect of the learners and teachers, and propagating state propaganda (Thobejane, 2013; Kallaway, 1988). Bantu education meant an inferior type of education that was designed to maintain the subordinate and marginal status of the majority racial group of the country. While the greatest divisions were between Bantu education for Blacks and an educational system for Whites, there were additional intermediate hierarchies in the educational system and general society. There was an education department that administered and regulated education for each ethnic and racial group separately (Thobejane, 2013). Four racial groups were recognized by the regime of the day: Black Africans were lumped together under the title "Bantu" by the nationalist government and later segregated along ethnic lines such as Zulus, Pedis, Tswanas, Sothos, Shangaan, Venda, and Ndebele while people of Indian and Asian descent were designated as "Indians". People of mixed descent, such as African, European, and Malay, were grouped under the label "Colored". People of Indian, Asian and those of mixed races (Coloreds) also received a somewhat inferior education to Whites, but comparatively much better off than the Bantu education system made available to Blacks (Stals, 2001). The segregated educational system was maintained institutionally by having a different department of education for each of the four racial and ethnic groups (i.e., Blacks, Indians, Coloreds, and Whites). Among these, the majority Blacks were at the bottom of the classification ladder. Apartheid (or Bantu) education in South Africa perpetuated hierarchical views of society and fostered an ideological consciousness of superior-inferior, master-servant, and ruler-ruled structure among all groups in South Africa The apartheid schools were doing exactly what John Holt, a deschooler, claims, namely that schools sort and indoctrinate students instead of liberating them (Msila, 2007).

The CNE principles on education for the Africans were declared as a way of maintaining the black South Africans in a permanent state of political and economic subordination. The education system had been an obvious instrument of control to protect power and privilege. And not only had apartheid education separated white children from black children; it had also divided white children into separate camps (Hartshorne, 1988). The black learners did not only feel the damaging effects of the CNE. It was destructive to the white learner as well. It is this history of South African education that necessitated the introduction of Curriculum 2005, sometimes referred to as outcomes based education (OBE). Its formulation involved cooperation among various stakeholders. The OBE system has introduced new learning styles implying change from passive, rote learning to creative learning and problem solving through active participation in the learning process. The National Qualifications Framework (NQF) provided the structure for the new curriculum in South Africa. The NQF was intended to prevent learners from being trapped in any one learning situation by facilitating movement between different areas and levels of education and training. The introduction of OBE in 1997 was a move towards changing the past system.

The Revised National Curriculum Statement (RNCS)

The revised national curriculum informs a system that seeks to introduce egalitarian pedagogy in South African schools. The goals of this system are to create a new South African identity that encompasses critical consciousness, to transform South African society, to promote democracy and to magnify learner involvement in education. This system is based on the Constitution of the Republic of South Africa, Act 108 of 1996, which provides the basis for curriculum transformation in South Africa. Education and the curriculum have a crucial role to play in achieving the aims of developing the full potential of learners as citizens of a democratic South Africa (DoE 2002:1). The RNCS envisages teachers who are qualified, competent, dedicated and caring. Teachers are supposed to assume various roles and these include being mediators of learning, interpreters and designers of Learning Programmes and materials, leaders, administrators and managers, scholars, researchers and lifelong learners, community members, citizens and pastors, assessors and Learning Area or Phase Specialists (DoE 2002: 3). All these roles are crucial for nurturing a new identity in learners. Learners are also expected to embrace the new values in the education system.

The promotion of values is important not only for the sake of personal development but also to ensure that a national South African identity is built on values very different from those that underpinned apartheid education (DoE 2002: 3): The kind of learner that is envisaged is one who will be inspired by these values, and who will act in the interests of the society based on respect for democracy, equality, human dignity, life and social justice. The curriculum seeks to create a lifelong learner who is confident and independent, literate, numerate, multi-skilled, compassionate, with respect for the environment and the ability to participate in society as a critical and active citizen. The above roles of the teacher and the learners are different from those under apartheid education. In the previous system, education was a one-way process where only the adult (the teacher) was expected to "impart knowledge" to the otherwise less competent child. Education was defined as the intervention of an educator in the life of a child to guide the child on his way to adulthood (Stuart, Van Niekerk, McDonald and De Klerk, 1987: 4). These writers further state that the role of the teacher was to transmit life contents to the child. The child who did not accept the authority of the teacher would not become a proper adult, hence the child had to be guided to obedience (Landman, Killian, Swanepoel and Bodenstein 1982: 11).

Democratic Education

The RNCS represents a pedagogy that has committed itself to reinventing education in South Africa. It opposes the traditional separatist education, which was the reproduction of inequality. Learners and teachers need to be exposed to a system that embraces democracy in order to build an egalitarian society. Education is the key because it empowers citizens to exercise their democratic rights and shape their destiny (DoE 2001: 13). Democratic education also enables people to participate in public life, think critically and act in a responsible manner. Democracy and education as a subject of discussion has not been without problematic areas. Chubb and Moe (1990: 23) state that there are no panaceas in social policy. This is supported by Plank and Boyd (1994: 270), who point out that democratic governance, is not a panacea for the problems of an educational system. However, Gutman (1999: 76) points out that the professional responsibility of teachers is to uphold the principle of no repression by cultivating the capacity for democratic deliberation.

As mentioned above, the values of the present South African education system are rooted in the values of the Constitution of the Republic. These ten fundamental values are democracy; social justice; equality; non-racism and non-sexism; ubuntu (human dignity); an open society; accountability; rule of law; respect; and reconciliation. All these values are highlighted in the Manifesto drawn up by the Working Group on Values in Education (Ministry of Education 2001: 111). The new system strives to create an identity of South Africanisms. Through education, South Africans yearn to affirm common citizenship and a sense of pride through commonly held values.

Post Colonialism

Postcolonialism alludes to the investigation of the social inheritance of expansionism and dominion, concentrating on the human results of the control and misuse of colonized individuals and their properties. The name postcolonialism is demonstrated on postmodernism, with which it shares certain ideas and strategies, and might be thought of as a response to or takeoff from imperialism similarly postmodernism is a response to innovation. The uncertain term imperialism may allude either to an arrangement of government or to a belief system or world view basic that framework, by and large postcolonialism speaks to an ideological reaction to colonialist thought, as opposed to just depicting a framework that comes after expansionism. The term postcolonial studies might be favored therefore.

Postcolonialism incorporates a wide assortment of methodologies, and theoreticians may not generally concede to a typical arrangement of definitions. On a straightforward dimension, it might look for through anthropological investigation to manufacture a superior comprehension of pioneer life from the perspective of the colonized individuals, in view of the presumption that the pilgrim rulers are problematic storytellers. On a more profound dimension, postcolonialism analyzes the social and political power connections that support expansionism and neocolonialism, including the social, political and social accounts encompassing the colonizer and the colonized. This methodology may cover with contemporary history and basic hypothesis, and may likewise draw precedents from history, political theory, rationality, social science, humanities, and human geology (Tordoff, 2014).

The Crises in the South Africa's Education System

Education is a noteworthy part of each country. In South Africa, it has been lauded, energized and regarded. This is because of the way that numerous South African guardians originate from a recently burdened foundation and were never allowed access to a good education opportunity. Post-1994, pretty much every parent frantically needed their youngster to be in school. Justifiably so the estimation of education has never been exceptional comprehended than how it was in post-Apartheid South Africa. Actually Grade 1 is a major ordeal for guardians who didn't have a similar chance. This is as yet an incredible bliss to numerous South African guardians.

Be that as it may, for a few, the delight blurs away as their youngster advances through school. This is an immediate aftereffect of the difficulties looking in the instruction framework. A framework that regularly creates confused students. This may appear to be brutal however there is solid proof to help the announcement. In March 2016, 15% of South Africa's complete spending plan was spent on fundamental educational training that is about R213.7 billion (World Economic Forum, 2016). Notwithstanding the cash spent, South Africa's essential instruction framework was appraised 126th out of 138 nations in the World Economic Forum 2016 17th Global Competitiveness Report. While South Africa's advanced education and preparing was positioned 134th. South Africa's training framework performs more terrible than less fortunate countries, for example, Zimbabwe, Kenya and Swaziland (SACMEQ, 2017). This obviously demonstrates the instruction framework does not mirror the cash put into the framework. Notwithstanding this reality, the administration keeps on structure more schools. Also, insights S.A's 2013 general family unit study referred to instructor related issues as the best test confronting government funded schools. The Southern and Eastern African Consortium for Monitoring Education Quality (SACMEQ) found that 60% of Math's instructors, teaching students from evaluation 1 to review 6, neglected to breeze through tests for Math's at the evaluation level.

Educators and Their Challenges

South Africa likewise had the most elevated rate of educator non-attendance, when contrasted with the other 14 nations that partook in SACMEQ III in 2007. Territories, for example, Gauteng, Free State and the Western Cape have put resources into preparing to improve region training frameworks. Notwithstanding, regarding the nation all in all, this is the most recent data accessible. In the event that the design is to improve the training framework, at that point it obviously demonstrates that the cash isn't including any reasonable esteem. It's simply there on the grounds that it must be there.

According to Weinberg (2015) the office should investigate embracing the educational modules and models utilized by autonomous schools. Or on the other hand in any event, they ought to explore cooperating. It is a well-known fact that autonomous schools exceed expectations, despite the fact that they just make up about 5% of the school-populace in South Africa. Maybe, South Africa can likewise get a few hints from other African countries. How well does the office really know the training frameworks in Kenya, Zimbabwe and Swaziland? Rather than looking into their own framework, more research could be put towards contemplating African nations that have better training rankings. The division could re-open educator preparing organizations that were shut post-1994. Organize the instruction capability. Having elevated expectations for instructors is something that the nations with best positioned training frameworks empower. For example, in Finland, educators are relied upon to obtain a graduate degree in the subject they instruct before they can educate (Torchia, 2014).

WHAT ABOUT THE CHILDREN?

All the more along these lines, the division could concentrate on the students. This could include investigating making a framework that bolsters the kids who drop out as well as remain stuck in specific evaluations. Maybe there ought to likewise be a framework for the individuals who come up short. The education office could likewise construct a framework that will coordinate a connection among schools and guardians. Wilson (2001) indicated that a framework that will make locally situated projects that guardians can use so as to empower a perusing society. Libraries are insufficient if students invest the majority of their energy at home. Guardians should be increasingly included, and the office needs to connect with them. This is key in making a decent education space where educators are truly captivating with students and guardians are empowering a learning attitude (Torchia, 2014).

The South African education is in emergency. Despite a tremendous development of the quantity of individuals in education today, it is an open inquiry whether the training they are getting is of a nice standard. Most kids entering Grade One in some random year are probably not going to register, and a considerably littler extent will total their 12 years of tutoring with a decent imprint in science. In 2015, the Organization for Economic Co-task and Development (OECD) discharged a report positioning the training frameworks of 76 nations from around the globe. The rankings were dictated by inspecting how well understudies did in math's and science tests.

South Africa performed horridly of the 76 nations contemplated, the OECD said that the nation had the 75th most exceedingly awful training framework. The main nation that performed more awful was Ghana. Singapore was positioned as the nation with the best instruction framework on the planet. The others in the best five were likewise Asian nations: Hong Kong, South Korea, Japan, and Taiwan, balancing the main five. The main other African nations in the rankings were Morocco, Tunisia, and Botswana, which were positioned 73rd, 64th, and 70th individually. All isn't lost, in any case. As the OECD's study calls attention to, Singapore's instruction framework is currently viewed as the best of those nations positioned, yet it had elevated amounts of absence of education during the 1960s. With political will and diligent work, it is conceivable to make solid training frameworks. Improving training likewise has major monetary advantages. The OECD evaluated that if all South African 15-year old's got a base dimension of training, the nation's total national output (GDP) would be multiple times bigger than if instruction levels stayed static.

Vincent and Chikerema (2014) concurrently indicated that viewing South Africa just against its companions in southern and eastern Africa, the image is marginally better, yet not without cause for concern. South Africa is an individual from the Southern and Eastern African Consortium for Monitoring and Educational Quality (SACMEQ). The association is comprised of sixteen African training services, which cooperate to screen and assess levels and guidelines of instruction in their nations. The fourth SACMEQ report directed in 2013 inspected the accomplishment of students in perusing and math's. South Africa indicated enhancements in both these controls. In perusing, South Africa improved from ninth to eighth, while the nation's positioning in arithmetic went from ninth to sixth. The normal South African student's score in 2013 was additionally higher or equivalent to the SACMEQ normal score. Be that as it may, two warnings should concern any individual who thinks about South African instruction or about their kids' future in the nation's training framework. To start with, the scores of educators declined notably. South African instructors previously participated in SACMEQ in 2007. In that year, about 80% of instructors had 'basic perusing' aptitudes, contrasted with 32.9% in 2013. The normal math's score accomplished by instructors likewise dropped, however not as altogether (Crush and Ramachandran, 2014).

There are genuinely huge contrasts in how South African students perform, contingent upon which region they live in, and whether they go to a no-charge state school, an expense paying state school, or a free school. For instance, taking a gander at the TIMMS science score, the two best performing territories were the Western Cape and Gauteng, while the two least fortunate performing areas were North West and the Eastern Cape. It is nothing unexpected that Gauteng and the Western Cape are the nation's two most extravagant areas (as far as GDP per capita), while the Eastern Cape is the least fortunate. The score of students going to a free school for math's was 13% superior to those in feepaying government funded schools and about 40% superior to those in non-expense state schools. Students at expense paying state schools additionally had scores almost 25% superior to anything those going to schools without any charges (Claassens, 2014).

Apparently, students from more unfortunate foundations likewise fared more regrettable than their happier partners. Students with guardians who had some post-tutoring instruction had math's scores 11% higher than those whose guardians had finished optional tutoring just, and 17% superior to anything those whose guardians had just some dimension of elementary school training. The patterns were comparable with respect to science. Students improved on the off chance that they lived in family units with running water and flush toilets. Students additionally improved in the event that they didn't encounter customary tormenting, and in the event that they talked the language that the test was given in, at home (Torchia, 2014).

Separated commonly, just three areas performed superior to the national normal. In the Western Cape, 55% of students did not meet the benchmark (couldn't peruse for importance), contrasted with 68.5% in Gauteng, and 73.4% in the Free State. As noted by World Economic Forum (2016) the Western Cape and Gauteng are the nation's two most extravagant areas, with the Free State having the third most elevated per capita pay. An astounding 91% of students in Limpopo could not peruse for significance in 2016. Similarly as with the TIMMS examine, there are a few issues that ought to be hailed and which are cause for concern. Young ladies performed far superior on the test than young men it is evaluated that young ladies' perusing capacity is as much as an entire year in front of that of young men. School foundation assumed a job in how well students performed. Students in schools with a library scored by and large 16% higher than those in schools with no library. Worryingly, over 60% of South African schools don't have libraries. Students from prosperous foundations likewise showed improvement over those from less wealthy ones. Students from a well-to-do foundation had scores about 40% higher than the individuals who could be delegated financially burdened, or neither impeded nor princely (World Economic Forum, 2016).

Differentiated Schools and Its Impact on Learners

As noted before, quintile one schools have the most exceedingly bad instructive results and these are the schools which most of dark youngsters visit. As indicated by information from the National Income Dynamics Survey (NIDS), almost 40% of dark students went to a quintile one school, with just nine percent going to a quintile five school. On the other hand, the extent of white students going to a quintile five school was over 90%, with irrelevant extents going to schools in alternate quintiles. Plainly dark kids are the ones who experience the ill effects of poor tutoring that is given by the administration. Combined with the high dropout rate, our schools do little to plan students for the later phases of their instructive life. In elementary school, the extent of students that needs to rehash a year is genuinely low in 2015 under 10% of students in evaluations one to seven were rehashing their specific evaluation. The rate hops in secondary school, in any case. In 2015, some 12% of evaluation eights were rehashing their evaluation (Weinberg, 2015).

By evaluation ten, completely a fifth of students were repeaters. The quantity of students rehashing was lower in evaluation 12 (just shy of eight percent), yet this is likely an aftereffect of the individuals who were probably not going to pass matric having been urged to leave school or start going to other instructive organizations (SACMEQ, 2017). Indeed, even the individuals who figure out how to pass matric and pass all around ok to enter college regularly battle at the following dimension of their instruction profession. In 2017, the quantity of individuals composing matric was a little more than 530 000, of whom seventy five percent passed and about 30% passed all around ok to be confessed to consider a four year certification at college. Be that as it may, the extent of students who passed and the individuals who picked up a single man's pass has expanded altogether since 2008, which brings up issues about the standard of matric. In 2008, the matric disregard rate was simply 60%, while just a fifth of matric students did alright to enter college. A 20% expansion in the genuine quantities of students passing and an over 40% expansion in the numbers passing alright to go to college proposes two things either our training framework has appeared solid improvement in the course of recent years or gauges are falling. Proof from our colleges would recommend the last mentioned. Prof Jansen has additionally raised a warning with respect to poor college instruction. He contends that the low measures at school level have 'invaded' colleges, an improvement, as per him, that was inescapable (SACMEQ, 2017).

The South African Democratic Teachers' Union (SADTU) has, rather than ensuring the privileges of educators, encroached the privileges of students in the nation. Considerably priest of essential training Angie Motshekga has scrutinized SADTU, saying that the association (which is a partner of the ANC through its participation of the Congress of South African Trade Unions) caused a larger number of issues than it understood in certain pieces of the country. She additionally wept over the association's resistance to measures to improve instruction, its hostile methodology, illicit strikes, and its utilization of arrangement matters as 'negotiating concessions' to get its direction. Research directed by an assignment group delegated by the Department of Basic Education (DBE) found that SADTU had successfully 'caught' the tutoring framework in six areas (the Free State, the Northern Cape, and Western Cape being the exemptions). The association was not just selling posts and controlling arrangements, and the control of arrangements was not just around the arrangement of educators and principals, yet additionally of area chiefs (Torchia, 2014).

However, in spite of the poor results at schools and the slippery impact of SADTU, there are moves to remove control from school overseeing bodies (SGBs). The Draft Basic Education Amendments Bill looks to remove the power that SGBs need to choose and designate senior educators, for example, principals. SGBs may almost certainly select and name junior posts. SGBs will likewise never again have command over a school's affirmation arrangement, with this power lying with the Head of Department. As indicated Weinberg (2015), various associations (not simply SADTU) contend that issues around the arrangement of teachers are a result of SGBs. To be sure, the main association that did not recognize SGBs as an issue in arrangements. SGBs, contending that they ought to be helped to meet their potential, and that more ought to be done to create them, to guarantee those serving on them know about 'proficient parts of posts or with the complexities of school the executives'.

Crush and Ramachandran (2014) indicated that the issue of language is utilized as a political football. Schools use it as a way to stay selective, while the administration utilizes it as a bar to beat those it objects to. Giving force back to guardian's short notice to change over to a twofold medium establishment when it isn't for all intents and purposes conceivable to do as such. The Federation of Governing Bodies of South African Schools (Fedsas) likewise cautioned that Afrikaans was under danger in Gauteng. The online application framework, whereby guardians apply online for their preferred school, never again gave guardians a choice of picking a language of guidance. Jaco Deacon, the vice president official of the association, said this was an endeavor by Mr Lesufi (Member of Executive Council) for Education (MEC) for Gauteng Province) to change all Afrikaans-medium schools in Gauteng to English-medium ones. He said that the extent of Afrikaans schools in the area was at that point low, with just six percent instructing solely in the language. The issue of language ought not to be utilized to reject individuals from accepting an instruction, yet it appears that there is a plan against Afrikaans schools in the area. Individuals ought to be permitted to pick the language of guidance for their youngsters, as long as it isn't done exclusively to prohibit individuals of different races and social foundations. The privilege to one's language and social personality is revered in the Constitution and must be secured and giving individuals more school decision is one method for doing this (Claassens, 2014).

Mr Lesufi MEC (Member of Executive Council) for Education for Gauteng Province indicated that guardians need increasingly decision, both in where they send their kids and in how their schools are run. Albeit tuition based schools are frequently seen as being bastions of benefit, this isn't generally the situation. Proof from around the globe demonstrates that individuals of little methods or winning low salaries are set up to influence real forfeits to send their kids to schools which to give a decent training (Vincent and Chikerema, 2014). Guardians send their youngsters to expense paying schools regardless of whether the alternative exists of a school without any charges. It is on the grounds that in numerous pieces of the world tuition based schools perform superior to anything state schools. They are frequently run better, with littler class sizes and instructors who educate for more and are missing less. Chapman (2011) indicated that, in Kenya it was discovered that half of youngsters in Nairobi went to a charge paying tuition based school despite the fact that Free State tutoring was accessible. Some have recommended this is essentially on the grounds that there are insufficient state funded school spaces accessible to oblige all kids, particularly in high thickness ghettos, for example, Kibera. Notwithstanding, work done by American scientists found that guardians were sending their kids to non-public schools regardless of the high budgetary expenses. It was assessed that having one kid in an ease non-public school in Nairobi cost about 12% of the salary of the essential worker. Taking into account that respondents in the study had a normal of three youngsters, picking a low-charge tuition based school over a free state school is probably not going to be a choice such guardians mess with. Low-expense tuition based schools in Nairobi were seen as being of higher quality than state schools, just as having harder-working educators. The main advantage that state schools had over non-public schools was foundation and learning materials. It is lighting up that individuals – as a rule – still picked tuition based schools over state schools, demonstrating the significance of

The education system is at stake. Even when we look at the provinces, those provinces dominated by blacks like Limpopo and Eastern Cape are always at the bottom but Gauteng always top the others with the average pass rate of 70,6% for Limpopo with only 21% bachelors and in Eastern Cape with 58.1 percent with only 22% bachelor degrees. With these poor results, it can only mean that these children are already excluded from getting into the Universities and their chances of a better future becomes slim. This takes us to the perspective by Karl Marx about the issues on class. Children from poor background and unrecognised languages struggle to get to the top, while others have access to WIFI, good libraries, parents who can afford to pay for extra classes, who understand the language to assist the children with school work and those children with helpers at home who do not have to struggle to help parents with fetching water from taps, wood for cooking or even just helping with domestic work What does it means? If we take African development as a serious issue, we cannot continue to ignore education for the mass.

CONCLUSION

It should be noted that all education activities that occurred over time has somehow been influenced by colonialism however, the main aim seem to have been intended for equipping black people among other things with new religious values and formal education. However, looking at missionary education activities, they reduced illiteracy and fostered literacy and numeracy amongst the Black people in South Africa. While Bantu education oppressed, limit and discriminate against black peoples progress and excellence and OBE was able to promote having ideas, knowing how to make decisions and problem solving amongst black students to improve their day to day lives. The goals of this the RNCS system were to create a new South African identity that encompasses critical consciousness, to transform South African society, promote democracy and to magnify learner involvement in education. Lastly the democratic education empowers citizens to exercise their democratic rights and shape their destiny therefore it can be concluded that colonialism did play a significant role in the evolution and transformation of education in South Africa.

South African instruction is in emergency. Huge numbers of the schools have poor offices, and neglect to get ready kids for life at college or in the work environment. Shockingly, the individuals who experience the ill effects of poor results are black youngsters. This has real ramifications for change and unraveling issues, for example, joblessness and destitution in South Africa. For whatever length of time that the nation is loaded with poor instructive results, these issues will stay for a considerable length of time to come, with desperate ramifications for the eventual fate of South Africa. Giving guardians more noteworthy power over the school their kids attend, and how that school works, will result in better results. All things considered, if guardians can pick the ethos and language arrangement, and have a state in the arrangements of educators and principals, learning results will be improved. Likewise, presenting a support framework will likewise result in schools improving state schools will never again be capable just to depend on a hostage showcase. Guaranteeing that schools give superb training is one of the center structure obstructs in making South Africa an effective society. Without great schools, delivering youngsters prepared for college and the universe of work, the issues faced will stay obstinate. Giving guardians progressively decision and opportunity with regards to learning will be an essential initial phase in confronting contemporary issues, just as putting the terrible heritage of politically-sanctioned racial segregation behind.

FUTURE RESEARCH DIRECTIONS

Recommendations are made that there is a need for research-based examples from the African context that can offer authentic suggestions for policy makers of what an African teaching methodology might be. Recognition of and sensitivity to local knowledge, culture and lived experiences need to provide guidance regarding the way in which teaching and learning among Africans unfolds. Greater effort still needs to be invested into developing and applying appropriate methods for African Ideology and Pedagogy .The essential principles of an African worldview are the prerequisites for the ideological cornerstones of an African centered paradigm.

REFERENCES

Abbatt, F., & McMahon, R. (1993). *Teaching Health Care Workers*. Macmillan.

Allen, R. E. (Ed.). (1990). *The Concise Oxford Dictionary of Current English* (1st ed.). Clarendon Press.

Behr, A. L., & Ngubentombi, S. V. S. (1988). *Education in South Africa*. Academica.

Boyd, A., & Lenix-Hooker, C. J. (1992). Afrocentricism: Hype or history. *Library Journal, 18*(117), 46–49.

Boyd, W. L. (1999). Paradoxes of Educational Policy and Productivity. *Educational Policy, 13*(2), 227–250. doi:10.1177/0895904899013002001

Brain, J. B. (1990). *The Catholic Church in the Transvaal*. Missionary Oblates of Mary Immaculate.

Chubb, J., & Moe, T. (1990). *Politics, Markets and America's Schools*. Brookings Institution.

Classens, A. (2014). *Communal land, property rights and traditional leadership*. Paper presented at Wits Institute for Social and Economic Research.

Cockrell, A. (1996). Rainbow Jurisprudence. *South African Journal on Human Rights, 12*(1), 1–38. doi:10.1080/02587203.1996.11834900

Crush, J., & S Ramachandran, (2014). *Xenophobic violence in South Africa: Denialism,minimalism, realism*. Southern African Migration Project Policy Series 66. Cape Town: SAMP.

De Houwer, J., & Moors, A. (2013). What is learning? On the nature and merits of a functional definition of learning. *Psychonomic Bulletin & Review, 20*(4), 631–642. doi:10.375813423-013-0386-3 PMID:23359420

Department of Education. (2002). *Revised National Curriculum Statement Grades R-9 (Schools) Policy, Arts and Culture*. Government Printer.

Gade, C. (2012). What is ubuntu? Different interpretations among South Africans of African descent. *South African Journal of Philosophy, 31*(3), 484–504. doi:10.1080/02580136.2012.10751789

Gemma, B., Lesikar, O. N., & Yunus, R. (2003). *Education, indigenous knowledge and globalisation*. Aang Serian Community College.

Gibson, J., & McDonald, H. (2001). *Truth-yes, reconciliation-maybe: South Africans judge the truth and reconciliation process*. Institute for Justice and Reconciliation.

Giliomee, H., & Mbenga, B. (2007). *New history of South Africa*. Tafelberg.

Graybill, L. (2002). *Truth and reconciliation in South Africa*. Lynne Rienner.

Hartshorne, K. (1992). *Crisis and challenge: Black education 1910-1990*. Oxford University Press.

Idang, G. E. 2015. African culture and values. *African Culture and Values, 16*(2), 97-111.

Kallaway, P. (Ed.). (1990). *Apartheid and education: the education of black South Africans* (4th ed.). Ravan Press.

Kritzinger, J. J. (1988). *The South African context/or mission*. Lux Verbi.

Kruger, E. G. (Ed.). (1986). *Education: past, present and future*. Euro Publications.

Landman, W., Killian, C., Swanepoel, E., & Bodenstein, H. (1982). *Fundamental Pedagogics*. Juta.

Lengrand, P. (1970). *An introduction to lifelong education*. UNESCO.

Letsekha, T. (2013). Revisiting the debate on the Africanisation of higher education: An appeal for a conceptual shift. *The Independent Journal of Teaching and Learning, 8*, 5–18. https://www.edge.org/conversation/lera_boroditsky-how-does-our-language-shape-the-way-we-think

Lewis, A. (1992). *Individual and community needs as aspects of the education crisis. A historical educational overview* (Unpublished MEd dissertation). University of Stellenbosch.

Mabunda, D. C. (1995). *A historico-educational survey and evaluation of Swiss Missionary Education at Lemana* (Unpublished MEd dissertation). Unisa.

Mawasha, A. L. (1969). *From Missionary and provincial control to state and community control of Bantu education: an appraisal* (Unpublished MEd dissertation). Unisa.

Melan, C. G. (2004). *Storytelling in Teaching*. The Association for Psychological Science.

Meyer, A. (2012). Suid-Afrika se oerverlede. Geskiedenis van Suid-Afrika.

Onuoha, F. (2015). *Locating African Values in Twenty–First Century Economics*. Available online at: https://www.africanliberty.org/2015/09/09/onuoha-frank-locating-african-values-in-twenty-first-century-economics/

Pearson, S. A. G. (2012). Colonialism and the African experience. In V. B. Khapoya (Ed.), *The African experience* (pp. 99–138). Longman.

Roney, R. C. (1996). Storytelling in the Classroom: Some Theoretical Thoughts. *Storytelling World*, *9*, 7–9.

Senge, P. (1990). *The Fifth Discipline*. Doubleday Currency.

Serrat, O. (2008). *Storytelling*. Available online at: https://www.adb.org/sites/default/files/publication/27637/storytelling.pdf

Stals, T. (2001). *The South African Economy: An Evaluation*. http://www.resbank.co.za/address/1996/ad110796html

Stones, E. (1966). *An introduction to educational psychology*. Methuen.

Stuart, J. F., Van Niekerk, L. J., McDonald, M. E. W., & De Klerk, D. (1987). *Didactics*. Southern Books.

Thobejane, T. D. (2013). *History of Apartheid Education and the Problems of Reconstruction in South Africa Sociology Study*. Academic Press.

Thompson, L. (1990). *A history of South Africa*. Yale University Press.

Torchia, C. (2014, Jan. 27). What next for the rainbow nation? Nelson Mandela's death casts shadow over racially charged South Africans. *The National Post*.

Tordoff, W. (2002). *Government and politics in Africa*. Indiana University Press.

Tyack, D., & Cuban, L. (1995). *Tinkering toward utopia: a century of public schools reform*. Harvard University Press.

Valji, N. (2003). *Creating the nation: The rise of violent xenophobia in the new South Africa* (MA thesis). York University.

Venter, I. S. J., & Van Heerden, S. M. (1992). *The grounding of history of education: an introduction*. Butterworths.

Verster, T. L. (1986). *Education themes in time perspective. Part 2*. Butterworths.

Viljoen, R. (2001). Aboriginal Khoikhoi servants and their masters in colonial Swellendam, South Africa, 1745–1795. *Agricultural History*, *75*(1), 28–51. doi:10.1525/ah.2001.75.1.28

Vincent, C., & Chikerema, A. F. (2014). Indeginisation of democracy: Harnessing traditional leadership in promoting democratic values in Zimbabwe. *Journal of Power*, *2*(1), 67–78.

Young, C. (2012). *The postcolonial state in Africa: Fifty years of independence. 1960–2010*. The University of Wisconsin Press.

KEY TERMS AND DEFINITIONS

African Values: African values means the written and unwritten moral standards that are commonly held by African people in general.

Education: Refers to a socially organized and regulated process of inculcating certain values as part of socialization process.

Indigenous Knowledge: Denotes a comprehensive knowledge system that is used by local people in securing their livelihoods.

Storytelling: Designate a form of sharing knowledge or experience through narratives as part of learning exercise.

Teaching and Learning: Mean a process whereby social values are inculcated to a learner of which form part of socialization programme.

Chapter 8
Theorising the Politics of Knowledge Production in Curriculum in Zimbabwe:
Indigenous Knowledge Systems for Transformative Classroom Practices

Nathan Moyo
Great Zimbabwe University, Zimbabwe

Jairos Gonye
Great Zimbabwe University, Zimbabwe

ABSTRACT

This chapter theorises the politics of knowledge production in order to understand the ways in which Indigenous Knowledge Systems (IKS) could be framed as bases for promoting transformative classroom practices in Zimbabwe. Doing so is necessary as the school curricula of many education systems in postcolonial Africa remain subservient to the Western European epistemology. The trope, transformative uncolonial learning, is employed in order to re-imagine an ethical pedagogy that could result in transformative classroom practices. The argument developed is that history and dance, as implicated in the politics of the black body, could be re-framed as the basis of ethical classroom practices. To achieve this, teachers need to embrace productive pedagogies that promote pluriversality of knowledges as valid and legitimate school knowledge.

DOI: 10.4018/978-1-7998-1249-4.ch008

INTRODUCTION

This chapter employs two school subjects, History and Dance, to theorise the ways in which Indigenous Knowledge Systems (IKS) could be framed as a basis for promoting productive pedagogies that have transformative potential to dismantle the dominant colonial underpinnings that inform the school curricula in many postcolonial states, Zimbabwe included. The modern nation of Zimbabwe was formerly known as Rhodesia, a colony of the British Empire. Colonisation forcibly incorporated the indigenous Africans into the capitalist system of production. To ensure unopposed exploitation of resources, colonialists harnessed the education system, and in the process, undermined the indigenous Zimbabwean knowledge systems. Consequently, a Eurocentric education system was imposed on the indigenous Zimbabweans who found themselves alienated from their history and cultural practices. Denied of such embodied anchoring, most black Zimbabwean recipients of Western European education found themselves willing participants in the denial and denigration of indigenous heritage, through an alien education system.

It is a sad reality that more than six decades after the first African states attained independence, the school curricula of many education systems in postcolonial Africa remain subservient to the Western European hegemonic epistemology (Breidlid, 2013). Regrettably, attempts by most postcolonial states at reforming their education systems have been largely futile (Bryant, 2018). Postcolonial curricula practices, grounded in Western epistemology as they are, have been inimical to the development of a critical and empowering educational paradigm. This reflects an ideological carryover of modernity, which, through the banner of "science", entrenches "the utilisation of [the] dominant western world view of knowing and knowledge production as the only way of knowing" (Kaya & Seleti, 2013, p.33). Indigenous knowledges such as those embedded in History and Dance are thus peripherised and subjected to epistemic violence (Kincheloe (2007). Such practices, unfortunately, perpetuate the mental subjugation of postcolonial learners. Breidlid (2013, p. 5) expresses the point graphically where he observes that, "the global architecture of education is hegemonic", which signifies that curricula discourses and practices worldwide misleadingly reflect imperialist epistemologies as the dominant grammars of learning. This not only results in undemocratic and unethical learning but also denies learners the opportunity to access the essential "transformative uncolonial learning" (Wane & Simmons, 2011, p. 3; Freire, 1990). This latter educational imperative could also enable the assertion of Indigenous knowledges and the reclamation of the African 'being' in an otherwise Eurocentric curriculum.

This chapter thus explores how postcolonial subjects, being trapped within a legacy of colonialism might begin to envisage epistemic insurrectionary strategies that could free them from the bondages of Eurocentric epistemologies (Gonye &

Moyo, 2018; Medina, 2011). Indeed, colonial education proved to be "epistemological fascism" (Paraskeva, 2016, p. 3), since it was "essentially disruptive, inadequate, and totally unsuited for African needs" (Bryant, 2018, p. 292). The chapter is thus located within the sociological/existentialist construct of the politics of knowledge production (Leonardo 2018; Breidlid, 2013; Darder, 2016; Sartre, 1973; Fanon, 1952/2008) in order to contest how the black body and indigenous knowledge, respectively, have been framed as invisible and unworthy of learning. This construct could enable an understanding of how knowledge is produced and reproduced, and its implications for the black body and its presence in the curriculum. In this regard, Wane (2008, p.184) poses this pertinent question, "Who controls knowledge and whose knowledge is considered valid?"

It is against this backdrop that the chapter employs the "transformative uncolonial learning" trope in order to reimagine an ethical pedagogy which seeks to "resurrect Indigenous knowledges [...], both on the micro and macro level, by claiming that indigenous knowledges have important assets that need to be seriously considered in a world that is completely dominated by Western epistemology and knowledge production" (Breidlid, 2013, p. 3). Here, we attempt to theorise how these assets as obtainable in History and Dance, both of which are repositories and media of IKS, could be the basis of productive and ethical pedagogies. History and Dance also share the body as a common factor, the latter of which has been 'forced' out of existence through racist repression and erasure. The primary research questions that the chapter addresses are, p.

- How can the History and Dance curricula be reimagined as transformative uncolonial learning on the basis of Indigenous knowledges?
- What pedagogical practices have potential for an ethical pedagogical practice in a typical postcolonial Zimbabwean classroom?

The focus is on History and Dance curricula since both subjects are implicated in the epistemological web of knowledge production. Both History and Dance embody and transcribe the lived experiences of the indigenes of a place. It was, therefore, no accident that the colonial epistemological onslaught targeted these two learning areas in the most vicious manner (Gonye & Moyo, 2015; Asante, 2000). In colonial Zimbabwe, the superimposition of the Western educational system on the African educational system facilitated the relegation of African traditional dance to the status of primitivism and paganism, and hence its marginalisation in the Western education curriculum (Gonye & Moyo, 2018; Gonye, 2013). For instance, Trevor-Roper (1964) describes 'dance-performing' Africans as a people without a history even as late as the 1960s.

Since the Western European nations considered themselves the centre of positive, rational sciences and a model of progress (Shizha, 2013) all their colonies would be subjected to curricula whose cultural constructions were informed by dominant Eurocentric worldviews, practices and pedagogy. This was despite the fact that the diverse pre-colonial African societies had viable indigenous education systems (Shizha, 2013). According to Amegago (2011, p. xiii) Africans boasted of a contextually-based and socially ethical "traditional African curriculum" where "[T] he arts were integral […,] and were transmitted through practice involving social interaction and a holistic integration of physical, emotional, spiritual, and intellectual values." However, the Western Europeans reconfigured other people's knowledge sources and systems including history and dance knowledges in binary terms, as inferior, primitive, irrational and retrogressive (Amegago, 2011; Shizha, 2013), while theirs were superior, modern, rational and progressive (Nkomo, 2011). This buttressed the Eurocentric notions of the indispensability of their knowledge systems which had to be imposed on the subordinated. In the postcolonial Zimbabwean Curriculum Framework, 2015-2022, both History and Dance Education are, however, foregrounded as critical learning areas embodying critical decolonising and emancipatory potency to challenge notions of racial superiority, individualism and greed that are the hallmark of Western epistemology.

The Politics of Knowledge and the Politics of the Black Body

The politics of knowledge production and reproduction in postcolonial curriculum theorising needs to be negotiated in the context and matrix of the coloniality of power. The coloniality of power is one which promotes pedagogical practices that work to invisiblise other knowledges, especially Indigenous knowledges (Ndlovu-Gatsheni, 2013, p. 8). Africa is presently caught up in the constraining global colonial matrices of power in which the African being's experience has been that of denial and aggression at the hands of slavery, colonialism/globalisation and the vestiges of a Eurocentric curriculum. Slavery, colonialism and Christianity have been instrumental in framing the people of the South, particularly Blacks, as "primitive" "backward" and "ahistorical" and whose practices are not worth preserving (Funez-Flores & Phillion, 2019). This has had a negative and telling effect on subjects such as African History and Dance.

To appreciate the effect of such Eurocentric framing on History and Dance, the chapter theorises the black body, particularly the dancing one as a defiant survivor. It does this through drawing from the lived experiences of one of the authors and an evaluation and extension of both their critical works (Gonye, 2013; 2015; Gonye & Moyo, 2018). It is widely recorded that the Whites/colonisers objectified the black body to deny it an existential presence, hence being (Amegago, 2011; Gottschild,

2003; Kuwor, 2013). Gottschild, (2003, p. 2) succinctly captures this global experience where she notes "The black dancing body was the negative space around which the white dancing body was configured." Since the black body was considered an anatomical "anomaly" (Gottschild, 2003, p. 3) it could not be selected for dance scholarship opportunities alongside white bodies. While this fate befell the 1950s African-American, it was even worse for the conquered African counterpart dancing on the 'dark continent'. For colonial politics to succeed, the 'animalistic' black body had to be tamed and dwarfed, and in that process, dehumanised and denied of any history. Stripped of being-ness in that world it [the black body] should be found, the black African body could claim neither a history nor a humane dance. There was also a coterminous denial of the epistemological value of IKS and its practices. Apparently erased out of existence, the politics of the black body remained alive, agitating for liberation and acknowledgement. Gittens (2012, p. 50) cites Dr Pearl Primus who claims that dance is a weapon to counter expressively the frustrations of "all [the] human beings who because of race, creed, or color, are invisible." Similarly, Gonye (2013) narrates graphically how the British colonialists' incarceration and attempted conversion to Christianity of Mbuya Nehanda, Zimbabwe's iconic 1890s First Chimurenga anti-colonial fighter, and her eventual execution, was a symbolic move to physically, politically, historically and culturally subdue the resisting black body. However, as Gonye (2013) recounts, the events leading to Nehanda's execution are equally symbolic. Mbuya Nehanda refuses to be converted to the coloniser's religion and, instead, performs a dance of defiance before her tormentors, an act which resonates with Africana existential philosophy. Dismissive of imminent execution, she utters an inspiring prophecy, p. *My bones shall rise again.* It can be argued that Nehanda's bones 'rise' metaphorically in the form of the new black bodies of the freedom fighters of the 1970s Second Chimurenga or Zimbabwean liberation struggle. The Black Zimbabwean body becomes re-inscribed with the politics of resistance and defiance as signified in the wartime dance - *kongonya*. This echoes Gonye & Moyo's (2018, p. 162) observation that "the body becomes both a symbolic and literal site for political subjugation and contestation." In short, the Zimbabwean experiences speak that the black body does exist and does move, meaning to say, through that movement and performance the body thereby communicates and passes history, not only about the context of performance but the very being of a people – which deserves to be learnt as valid knowledge. As Kuwor (2013, p. ii) avers, dance is a "dynamic tool in the transmission of indigenous knowledge, skills, values and virtues." However, as much as this suggests that defiant dancers were communicating decolonial history and culture to the participants and watchers, nothing seemed to have changed in the postcolonial curriculum, the latter of which seemed to be under the perpetual grip of the coloniality of knowledge.

The coloniality of knowledge requires that postcolonial scholars engage in an epistemic insurrection (Gonye & Moyo, 2018; Medina 2011) that dismantles the Western epistemological stranglehold. Such a move is predicated on Mignolo's (2009) epistemic disobedience, which reverberates with "a double gesture --- [that... first,] calls for the re-embodiment and relocation of thought in order to unmask the limited situation of modern epistemologies and their linkages with coloniality" and, "second, [that] demands delinking oneself from these knowledge systems and reimagining present-futures" (Desai and Sanya, 2016, p. 12). A double-edged heuristic such as this is critical in challenging the apparent silences of Western epistemologies concerning the knowledge systems of the formerly colonised, as well as the excavation of those silences, and affirmation of the epistemic rights of the margins (Desai and Sanya, 2016). The argument herein is anchored on black identitarian combativeness in both thinking and praxis of Africana existential philosophy. According to Sithole (2016, p. 178) "Africana existential philosophy rests on the lived experiences engaging in the epistemic rupture of silenced and marginalised discourses, which are concerned with fighting epistemological violence and accounting for the subjectivity of the black subject"

It is in pursuit of a curriculum of indigenous Zimbabwean knowledge, therefore, that it becomes a political and pedagogical imperative that we re-ask the foundational questions about what counts as knowledge and whose knowledge counts in an historical context (Kelly, Luke & Green, 2007). Africana/black existential theory resonates with critical race theory (CRT), especially regarding the refutation of racist constructions of the African body and IKS as shown in the CRT tenets of "experiential knowledge" and the desire to "challenge dominant ideologies" for "social justice" (Yosso, 2005, pp. 72–75) for those who have not been considered as human (Bassey, 2007). This current chapter extends Gonye & Moyo's re-adaptation of Solórzano and Yosso's (2002) pertinent question in counter-storytelling to read, p. Whose history and dances are privileged in the arts curriculum and whose history and dances are under-represented and marginalized? (Gonye & Moyo, 2018, p. 161). These political and ethical questions resonate with urgent calls for the recovery of multifaceted forms of IKS. For Odora-Hoppers and Makhale-Mahlangu (1998, pp. 8–9), for instance, IKS combine different knowledge systems that "encompass[ing] technology, social, economic and philosophical learning [...]. It is knowledge relating to technological, social, institutional, scientific and developmental systems including those used in the liberation struggle." It is therefore important to ask how to reconceptualise those Indigenous knowledges that have been historically marginalised and how to centralise them in transformative and empowering curriculum practices.

It is against this backdrop to reimagine inclusive, democratic and socially empowering curricula that since independence in 1980, post-independence Zimbabwe has embarked on major curriculum reform initiatives in an attempt to recalibrate the ideological underpinnings of the inherited curriculum. The apogee of this epistemological struggle was the 2017 promulgation of the Zimbabwe Curriculum Framework for Primary and Secondary Education, 2015-2022. Before then, in 1999, almost two decades after the attainment of independence, the Commission of Inquiry into Education and Training (CIET) had noted that Zimbabwe lacked a clearly articulated philosophy of education as the socialist rhetoric of the 1980s had become irrelevant following the collapse of Socialism and the subsequent adoption of the neoliberal Economic Structural Adjustment programme of the 1990s. CIET had thus proposed the adoption of *Ubuntu/Unhu* as the philosophical underpinning of the education. Significantly, this recommendation has been adopted in the Curriculum Framework, 2015-2022.

The findings below seek to address the questions on how History and Dance curricula could be reimagined as transformative uncolonial learning on the basis of Indigenous knowledges and on which pedagogical practices could best enhance ethical epistemic practices in a typical postcolonial classroom.

FINDINGS AND DISCUSSION

The Zimbabwe Curriculum Framework for Primary and Secondary Education, 2015-2022, attempts to foreground participatory and empowering pedagogies to counter the authoritarian colonial/missionary educational practices that sought to pacify the indigenous learners. Following the establishment of the missionaries' spearheaded education system, the Indigenous Zimbabwean learners were incorporated into the Western epistemological empire and knowledge production. Missionaries were keen to suppress so-called pagan sources of knowledge such as African traditional dance (Gonye & Moyo, 2015). Windel (2009, p. 4) writes that "Education was a crucial point of interaction between 'the native' and the settler." European missions had begun to independently set up schools that would enrol indigenous learners. The Foucauldian adage, 'Knowledge is Power' (Foucault 1980) was routinely brandished among the colonised indigenes who found themselves both dazzled and attracted by the new 'wisdom'. This subtle incorporation constituted epistemic violence. Indigenous knowledges were classified as 'without power' and, therefore, worthless knowledge. As Ndlovu-Gatsheni (2015, p. 15) puts it, "Africa was problematically and forcibly integrated into the evolving Euro-North American-centric modernity and its capitalist system over the last 500 years." Significantly, during this period, "African history was denied or appropriated; African culture [was] belittled; the status and standing of Africans as human beings was called into question" (Ahluwalia & Nursey-Bray 1997, p. 2). Regarding African dance, eminent Eurocentric writers invented terms and expressions to disparage dance practices with the intention of making Africans feel ashamed and estranged from their practices, which, ironically, embody their knowledges. Trevor-Roper (1964) would denigratingly claim that of all the rudimentary knowledges Africans had, the basest was the propensity to gyrate in some dark corners, towards which the European was to bring enlightenment (Gonye, 2015).

To achieve its mission, the colonial administration crafted the 'Native Education Policy' which became a major front in the battle of competing interests among the colonial state, missionary societies, international organizations and governing bodies, and Africans themselves (Windel, 2009). The policy facilitated epistemic violence on indigenous knowledges. In Rhodesia, for instance, a series of education Acts were passed, *viz*, p. the 1899 Education Ordinance; 1903 Education Ordinance; 1907 Education Ordinance; 1929 Department of Native Development Act; 1930 Compulsory Education Act; 1959 African Education Act; 1973 Education Act; and the 1979 Education Act (Richards & Govere, 2003). These Acts created a form of 'educational apartheid' where separate and unequal education systems for the Indigenous people and settlers were developed on the basis of race. The 1899 Education Act laid the foundations for an obnoxious system of education for the Africans as it provided "the ways and means of government control over the content of what was taught to African students, and even what the purpose was for educating Africans," (Richards and Govere, 2003, p. 139). Inevitably, subjects that were deemed to have 'subversive' potential, such as Indigenous History and Dance were deliberately excluded in order to "forestall political movements that threatened the colonial order" (Windel, 2009, p. 2). Instead, they promoted the so-called 'activity curriculum' which emphasised agricultural and manual skills. Such practices ensured the subordination of the Indigenes in epistemological and cultural ways.

The Curriculum Framework, 2015-2022 is a significant landmark in the political and educational landscape of Zimbabwe. It articulates a consolidated view of the nation's school system. Hitherto, piece meal measures on an *ad hoc* basis had been employed to frame the policy direction of education in Zimbabwe. The Framework is predicated on the country's 2013 Constitution, an outcome of democratic processes obtaining in the country then, unlike the 1979 Lancaster House Constitution which had been imposed on the people to pave way to independence. The Framework specifically draws from Articles 16 and 33, which, respectively, advocate the "upholding [of] our cultural heritage" and the "preservation of [Zimbabwe's] traditional knowledge" among others (MoPSE, 2015, p. 14). This means that this decolonial and reclaim-indigenous-knowledge thrust in the curriculum, which we believe is embodied in subjects such as Dance and History is actually enshrined as a constitutional right. The Framework was preceded by widespread consultation of the key stakeholders, which suggests maturing democratic practices in the country.

The above factors provided an ideal moment for the articulation of an all-encompassing educational policy that would be sensitive to the nation's past, and aspirations for the future. At the same time, the Framework was drafted in full cognisance of the worldwide changes in educational theory and practice in the last three decades and the consequent shifts in educational thinking. These shifts which have a direct bearing on both curriculum, teaching and learning include, p. reconceiving of curriculum as product to curriculum as both process and product; reimagining teaching as not transfer of facts but as learner construction of knowledge, and pedagogy, as not didactic teaching, but participatory, learner-centred and interactive methodologies (MoPSE, 2015, p. 10). The Framework thus undertakes a delicate act of valuing the disciplinary knowledge inherent in the traditional subjects and infusing it with the Indigenous knowledges in pursuit of learning as an engaged lifelong activity. As stated in the foreword, "Education should mould learners who cherish and practise the Zimbabwean philosophical orientation of *Unhu/Ubuntu/Vumunhu,"* (MoPSE, 2015, p. iv). The political will to back these changes is captured in the Preamble by the President of the Republic thus, p.

Over the years I have called for relevant, quality and inclusive education in the school curriculum whose hallmarks are competences desired in life and work. Simultaneously, such education must remain accessible and affordable (MoPSE, 2015, p. i)

The above remark underscores the need to "provide a transformative educational framework," (MoPSE, 2015, p. iii) in pursuit of the nation's socio-economic trajectory. It is also indicative of how political rhetoric has been deployed to support educational reform in the country. Significantly, the fiery rhetoric characteristic of the leadership as displayed in other fora is here toned down, thus making the document a pragmatic effort. Taking its cue from such a pragmatic stance, the Framework then articulates the fundamental aim of the curriculum as "to produce a highly competent learner who has a Zimbabwean outlook" (MoPSE, 2015, p. iii) without at the same time invalidating knowledges that are foregrounded by global modernity. The focus is on enhancing the relevance of education through fostering, p.

- Patriotism
- Awareness of heritage, history, culture and traditions
- Inter-cultural understanding and tolerance
- Self-respect and respect for others(*Ubuntu/Unhu/Vumunhu*)
- Being open, receptive, assertive and optimistic

In order to ensure that learners are thoroughly grounded in their Zimbabwean heritage, history, culture and traditions, the Framework then explicates that "Learners will be supported in developing and promoting their ethnic identity and stressing the linkages between local cultures and national identity" all of which will be "promoted through learning about each other's arts and crafts, music and dances, poems, rituals, cuisine and apparel rooted in Indigenous Knowledge Systems (IKS)" (MoPSE, 2015). Herein, emerge both History and Dance as critical learning areas alongside music, poems and cuisine. And such learning has to be promoted through ethical pedagogical practices that are consonant with the tenets of IKS. These cognate learning areas of History and Dance constitute the embedded knowledges of the indigenous people. Recentring these subjects as critical to learning in the new curriculum is thus a recognition of how the coloniality of knowledge needs to be disrupted through foregrounding IKS, which challenges "the very core of knowledge production and purpose" (Dei, 2014, p.173).

The Framework elaborates its unequivocal epistemological stance on IKS when it states, as regards acquisition of competencies in Environmental education, that, p. "As learners acquire these competencies they will be informed by both Indigenous Knowledge and other Knowledge Systems" (MoPSE, 2015, p.7). Here, the Framework repositions IKS as valid and legitimate knowledge for understanding one's environment, alongside other knowledges that are Eurocentric. In foregrounding such pluriversality of knowledges as valid and legitimate school knowledge, the Framework is revolutionary in that the coloniality of knowledge is debunked. At the same time, the call for complementarity of knowledges is indicative of Bhabha's (1994) hybridity and transdisciplinary as hallmarks of holistic and integrated learning. Such an approach resonates with Dei's (2014, p.174) assertion that, "embedded in such knowledge is an understanding that society, nature, and culture interface", and the particular responsibilities required of African peoples in maintaining and sustaining their local environments. Thus, the updated Framework marks those incipient steps towards a transformative uncolonial learning as ethical classroom practice.

The definition of curriculum that the Framework adopts is significant for the politics of knowledge construction that could be at play in the pedagogical transactions that are envisaged. The Framework initially represents curriculum as:

the sum total of all learning experiences and opportunities that are provided to learners in the context of formal and non-formal education (MoPSE, 2015, p.2).

This could be described as a modernist definition that falls within the scope of Jackson's (1968) narrow definitions of curriculum. Such a definition ignores the critical questions about knowledge production and reproduction. It seemingly ignores how learners, particularly indigenous Zimbabwean learners, are positioned by such a notion of curriculum. However, in Section 1.6.4 the Framework moves beyond this narrow definition where it states that, p.

This curriculum integrates the values, principles and practices of participatory action based on the equal access to, and exercise of, rights, duties and responsibilities in the context of families, communities and stakeholders. Tolerance and respect are at the heart of management of diversity, differences and conflicts (MoPSE, 2015, p.4)

This elaboration is significant because it invokes the notion that curriculum is a political text which resonates with the lived realities of those who study it. Commenting on the political and value-laden-ness of curriculum Dei (2014, p.178) notes that, "the curriculum is power-saturated and involves the power to construct, validate, and legitimize knowledge, and what is acceptable and not". The current participatory and action-based pedagogical approaches are a complete departure from the authoritarian methods of the colonial classroom.

In its reframing of education, the Framework posits that "The education system must focus on mastery of content knowledge and at the same time meaningfully relate that knowledge to practice" (MoPSE, p. 18). Consequent to this, the new curriculum posits, p.

The Zimbabwe Curriculum Framework promotes a competency-based approach which is realised through practical-oriented learning. It is envisaged that the curriculum shifts from being content-based (examination bound) to a competency-based (outcomes oriented) curriculum which focuses on the learners' capacity to apply knowledge, skills and attitudes in an independent, practical and responsible way (MoPSE, 2015, p.4)

This is an attempt to negotiate the binaries, theory and practice, a banality characteristic of colonial education. On the contrary, Indigenous education has always been a combination of theory and practice. Nyerere (1982, p.236) describes such education as "learning by living and doing" in ways that made the education more directly relevant to the society in which the child was growing up.

Another innovation that has potential for transformative uncolonial learning is where the Framework clarifies the underpinning curriculum principles. These are enhanced through cross cutting themes to be taught and learnt in context. Table 1 summarises the principles and the cross-cutting themes.

Table 1. Principles and cross-cutting themes in the Framework

Principles	Cross Cutting Themes
Inclusivity	Environment
Life-long learning	Health education
Equity and fairness	Heritage studies
Gender sensitivity	Enterprise and financial literacy
Respect (Ubuntu/Unhu/Vumunhu)	
Responsiveness	
Balance	
Diversity	
Transparency	
Accountability	

These principles resonate with the underlying themes of Indigenous education that existed prior to the introduction of formal schooling. Breidlid (2013) bemoans how colonialism sought to deny epistemic diversity while fostering inferiority in the indigene. Similarly, life-long learning, respect and care for the environment were central to the social fabric of Indigenous societies. Their inclusion here is a hankering for a society uncorrupted by Western decadence. The cognate disciplines of History and Dance have potential to reinvigorate these cherished values when they are reconceptualised through an indigenous lens. In the following section the chapter examines the kind of pedagogy that the Curriculum Framework foregrounds.

How pedagogy is framed is critical in understanding the assumptions that underpin subject content and how learners are positioned *viz-a-vis* the content that they are exposed to. This is because "pedagogy re-mediates, frames and rearticulates what will count as knowledge in classrooms" (Wrigely, 2003, p.92). The Curriculum Framework is unequivocal in addressing pedagogy. It states that learner-centred methods of teaching and learning should be the bedrock of all curriculum practice. The focus in the classroom should be on "knowledge generation not just information delivery and treating learners as empty vessels ready to be filled with information" (MoPSE, 2015, p.41). The imperative for knowledge generation is acknowledgement of the politics of knowledge production and that curricular knowledge is not a given but a co-construction undertaken by the learners and the teachers in the context of their educational milieu. While this renditions the Freirean pedagogy (1990), it also re-echoes the participatory learning approaches extant in Indigenous epistemological practices. Such a pedagogical approach is both empowering and liberating. It recognises the contingency of knowledge and affirms learners' abilities and agency in invoking their lived experiences as central to learning. Most importantly for this chapter, such recognition is at the root of an ethical curriculum practice. The Framework elaborates that:

The focus on learning revolves around learners as they engage in the search and discovery of new knowledge. The teacher acts as a co-explorer and facilitator in knowledge discovery in order to arrive at an objective understanding of content and demonstration of skills so acquired (MoPSE, 2015, p.41).

In pursuit of the above aims, the Framework posits that the inquiry-based approach is one of the pedagogical strategies to be used. This approach is aimed at "nurturing thinking, reflection and problem-solving among learners" (MoPSE, 2015, p.41). Thus, learning in the classroom will proceed from "lower to higher levels of cognition, with learners expected to work collaboratively with classmates and others beyond the school" (MoPSE, 2015, p.41). In addition, the learning is expected to be research-driven, interdisciplinary and involving performances, projects and multiple forms of media. These methods offer learners the opportunity to reconstruct their own set of meanings and understandings of the world which may act as a transformative uncolonial learning. The teacher is rethought of as facilitator-coach in ways that prod learners as active creators. Both History and Dance could offer exciting opportunities for a transformative uncolonial learning as will be explored below.

Decolonial Pedagogical Strategies in the History Curriculum in Zimbabwe

Ndlovu-Gatsheni (2015, p.16) underscores the salience of the History curriculum as a critical emancipatory pedagogy in the assertion that "theft of history for Africa translated into theft of its future." This theft is manifest in the ways in which colonial historiography attempted to dispute the historical fact that the Great Zimbabwe Monument was the work of the Indigenous Shona people (Garlake, 1987). Today, it is however acknowledged that Great Zimbabwe is, indeed, the work of the people of the Great Zimbabwe State which existed around 1450AD. This fact demonstrates an act of reclamation of the history by the Indigenous people, to which Falola (2005, p.503) attests; "if the 'colonial library' wanted to supress knowledge about Zimbabwe, national history rescued it." The reclamation of this history could be enhanced through decolonial pedagogical strategies that are grounded in transformative learning.

Such a bold move would in turn promote epistemological inclusion of Indigenous history, which Shanguhyia and Falola (2018, p. 4) advocate for where they note the urgent need to "furnish Africa and Africans with a historiography that would give purpose and meaning to former colonial subjects and their newfound sense of belonging."

Zimbabwean policy makers have attempted to address the above-stated need through successive curriculum innovations that sought to refocus the teaching of History from its Eurocentric concerns and linearity to an African, Zimbabwean and internationalist perspective grounded in critical disciplinarity (Moyo and Modiba, 2013; Moyo, 2014; Barnes, 2007). Thus, the Zimbabwe Curriculum Framework for Primary and Secondary Education, 2015-2022, cognisant of some of the challenges in the teaching and learning of History as identified by Chung (2008), promulgated a new History Syllabus F1-4. *The History Syllabus Forms 1-4, p. 2015-2022* (CDUTS, 2015) builds on the attempts of earlier post-independence syllabi to promote a typical Zimbabwean historiography as stated in the introduction, p.

The syllabus helps them (learners) to acquire an informed and critical understanding of social, economic and political historical developments. --- In addition, it fosters an understanding and appreciation of issues concerning population, human rights, democracy, empowerment, gender sensitivity, Unhu/Ubuntu/ Vumunhu (societal norms and values) and instils patriotism, leading to a change in learners' attitudes and behaviour.

Significantly, critical understanding of History is stated as paramount in the new syllabus, unlike the rote memorisation of facts and dates that was a legacy of classical Western education. Such rote memorisation had been reincarnated in the previous Syllabus 2167. The *Unhu/Ubuntu/ Vumunhu* philosophy is highlighted as the ideological underpinning of the History curriculum. This is a conscious attempt to enhance African-centred interpretation of history in ways that counteract Eurocentric interpretations that had been essentialised in previous syllabi. This also constitutes a critical acknowledgement that curriculum and history learning are about values, ideas, practices, identities, and how knowledge production is linked to identities, power relations, and pedagogy (Dei 2014).

The new History syllabus refocuses the content of the History curriculum through the introduction of topics that focus on Zimbabwean and African History. Such a nationalist historiography is intended to draw learners' attention to African anticolonial struggles across East and Central Africa, themes that lasted into the 1980s. A typical example is how the grand narrative of the Chimurenga War has become canonised in Zimbabwean history and is the central motif around which the nation-state is imagined. However, among the topics that are now taught are, p.

- Conceptualisation of History
- Origins of Humankind

- Ancient Civilisation in Africa, p. Egypt
- Development of Zimbabwean Societies
- Slavery and Slave Trade
- Early European Contacts with Zimbabwe
- Colonisation

The above topics have replaced the Eurocentric curricula whose topics had focused on Europe and the Americas (Moyo, 2014; Moyana, 1989). The Eurocentric curriculum had glorified conquest of the country, with the hailing of Rhodes (from wherein the colonial name Rhodesia originates) as a hero, while the Ndebele King Lobengula was dismissed as insignificant. As elsewhere on the continent, the African precolonial past was grossly ignored in the teaching of history. However, a mere change of topics without a philosophical re-orientation of history that is informed by Indigenous knowledges is not likely to lead to transformative uncolonial learning. An ethical approach to the learning of History requires the engagement of productive pedagogies that speak to the pluriversality of knowledges. Informed by the politics of knowledge production, the syllabus needs to trouble the colonial matrix of knowledge by asking the question: What does this history curriculum want [learners] to be? (Moyo & Modiba, 2013). Moyo and Modiba's question is grounded in Ellsworth's (1997) modes of address as a means of reimagining how learners are positioned by the particular histories that they learn in school. Segall (1999, p.364) reminds us that "meanings given to the past are never objective or neutral; they are always positioned and positioning." It is therefore important that there be productive pedagogical strategies that could enable learners to question the sources of history in ways that could lead to transformative uncolonial learning. To this end, the *History Syllabus Forms 1-4, p. 2015-2022* suggests critical teaching approaches across all levels.

The inquiry approach that is foregrounded in the Framework permeates the *History Syllabus Forms 1-4, p. 2015-2022* as well. This makes the teaching and learning of History a problem-based subject and not mere recall of facts and dates. This inquiry model to school history enables learners to reframe the curriculum as a site for the reconstruction of meaning and not the perpetuation of taken-for-granted assumptions that resonate with the ideological proclivities of Western epistemology. The curriculum thus becomes a social practice caught up in the (re)production of shifting networks and formations of power, knowledge and desire, (Moyo, 2014). The pedagogical strategies that teachers could employ thus include the use of:

- Games and Quizzes
- Imitation or Simulation
- Video and film shows

- Educational tours
- Drama, song and poetry
- Research
- Project
- Resource person
- Case study
- Folklore, and
- E-learning

Some of the strategies suggested above have potential to enable learners to rethink how historical knowledge is produced and structured, thus sensitising them to the contingency of historical knowledge. The use of song, dance and poetry as a strategy opens doors to capture significant historical interpretations that are not necessarily captured in official history texts. Similarly, engaging in project work enables learners to construct their own local histories while also working as professional historians. The use of resource persons is likely to enable learners to appreciate that historical knowledge is embodied reality that is not confined to the textbook, while also re-echoing the role of the griots and story tellers in traditional African societies. Taken together, the above pedagogical strategies embody a critical stance which may help raise the political consciousness of both teachers and learners towards how they are positioned towards the dominant knowledge in the curriculum. This is a critical step towards transformative uncolonial learning which has potential for enabling students to critically engage and actively change their world. As the following section shows, the Dance Syllabus F1-4, too, has potential to enhance the kind of transformative uncolonial learning that unravels the coloniality of power.

DECOLONIAL PEDAGOGICAL STRATEGIES IN THE DANCE CURRICULUM IN ZIMBABWE

The development of African dance in colonial Rhodesia suffered from Eurocentric "reductionist classifications" that objectified and fetishized the "dances and dancers from Africa as sensual, primitive, and mere physical activities valued only as products for insatiable white gaze" (Mabingo, 2019, p.4). This was despite the fact African dances were culturally diverse and carried cultural communicative and educative value (Nyamwaka, Ondima, Kemoni & Maangi, 2013). Indeed, the performing body is an archive of a nation's history and hence source of decolonial education. Resultantly, the empowering and emancipatory indigenous knowledge systems embodied therein were marginalised (Banks, 2009). Indigenous Zimbabwean learners were thus denied the opportunity to benefit from indigenous forms and styles of learning. For example, if the benighted African child had to learn dance at all, it would not be the so-called pagan and licentious ones (Asante, 2000) such as *mbakumba, chinyambera* or *mbende/jerusarema* but the modern virtuous European ones; ballet, foxtrot or ballroom. This was a manifestation of the disorienting effects of the coloniality of power. Indeed, the *mbende/jerusarema* dance was forbidden for performance and practice across colonial Zimbabwe (Asante, 2000; Gonye, 2015).

The view that Western European education in the former colonies has been just as violent and disruptive as the colonial conquest itself is indisputable, being "a tool to dominate and discriminate across the globe" (Denomme-Welch & Montero, 2014, p.137). That being the case, exclusionist tendencies in colonial art education, including dance had to be contested. With that prerogative in mind, Sorgel (2011, p.84) cites Tiérou (2001) who argues that "[W]hile the colonialist negation of African humanity and culture persists in the current debates on dance – which more often than not reiterate stereotypical views on African dance as being tribal, primitive, sensual, and exotic" there is still room for Africans to theoretically and choreographically reconfigure their dances in more liberating ways. This could be possible with a curriculum framework such as the new Zimbabwean one that incorporates indigenous views and practices into its content and pedagogy. As Banks (2009) suggests, dance pedagogy has the potential to decolonise the subordinated body and spirit. Such potential for a decolonial pedagogy can be demonstrated through reference to the Zimbabwean Dance Syllabus *1-4, p. 2015-2022* which was launched as part of the Curriculum Framework 2014-2022 (CDUTS, 2015).

The Dance *Syllabus Forms 1-4, p. 2015-2022* (CDUTS, 2015) for Zimbabwean secondary schools was compiled in 2015 and implemented in line with *Ubuntu/Unhu/Vumunhu* inspired Curriculum Framework, 2015-2022. The *Unhu/Ubuntu/Vumunhu* philosophy "epitomises universal human inter-dependence, solidarity, humaneness and sense of community common in African societies" (MoPSE, 2015, p.13). This suggests the Framework's deliberate attempt not only to recentre Africanness but also to address the injustices of the anti-indigenous Western European colonial curriculum through an embracing of dances from across the world. Current trends, globally, consider dance as an "honest", "authentic", "organic" or "holistic" expression of feelings with a "subversive power and socio-critical and emancipatory potential, from which [dance] has derived and still derives its avant-garde claims as well as its educational mandate." (Klein, 2011, p.17). This means that dance is an honest and authentic representation of a nation's unravelling story and trajectory. Dance is, therefore, both an expressive embodiment of indigenous knowledges and a subject that humanistically empowers its learners, hence, a potent tool for decolonial and ethical pedagogical practices. In this regard, dance teaching and learning that foregrounds IKS must be a 'must-do subject' for those learners located in the Global south.

It is important to remember that through drawing from the IKS the dance subject in Zimbabwe could enable the learner to creatively, actively and productively contribute to the "economic, religious, political and social development" of the learner, the society and the country (MoPSE, 2015, p.1). However, in the Zimbabwean curriculum, dance remains, like in many other countries (Connell, 2009), a subject still in its infancy (Gonye & Moyo, 2015; 2018). It took some drastic reforms in late post-independence, culminating in the 2015 curriculum review, for dance to be weaned from music and allowed its decolonial free berth. Commenting on a similar reluctance in Sweden, Styrke (2015, p.204) notes that "In order for dance to find its form as a school subject, a transformation is necessary." This implies that the general perception of dance should be modified in a positive manner that recognizes the embodied significance of dance to the nation and the individual. This is an arduous task in formerly colonised nations such as Zimbabwe where citizens had been conned to shun their own indigenous knowledge systems and practices.

In this chapter, we thus theorize indigenous Zimbabwean dance as a participatory means through which the cultural and political past and present of the nation are expressively projected and reclaimed; the citizen's tastes and corporeal presence are expressed; it is both a representation of indigenous people's stable and dynamic value belief systems and the performing individual's potential; also that indigenous Zimbabwean dance communicates what the performed cultures have come to represent over the years and what the performers believe the contemporary generations may, too, cherish and wish to preserve or modify. It is, however, important to note that there are numerous and different types of indigenous dances that were traditionally performed in the country. These are important considerations to remember in the dance curriculum, cognisant of the fact that the meanings of indigenous dance shall not remain static and immutable, as culture is dynamic and its constituents, hybrid.

The Zimbabwean secondary school syllabus for dance consists of seven major topics to be broken down over the four years of junior to intermediate secondary school education. The syllabus' broad developmental topics include, p.

- History of Zimbabwean, African, African-Diasporan and Contemporary Dance;
- Dance Techniques;
- Creative Composition and Intellectual Property Law;
- Anatomy, Nutrition and Identity;
- Aesthetic Values and Performance Appreciation;
- Technology in Dance; and
- Dance Management (Enterprise Skills).

The syllabus assumes an inclusive approach where indigenous and world dances; pure and hybrid dances; local and global choreographies, among other issues, are studied. This is unlike the racially exclusive colonial curriculum. The syllabus, however, begins by situating both the subject and the learner in the local historico-cultural context before moving outward into the African continent, Diaspora and the world at large. This is an ethical and socially just approach that neither seeks to estrange the learner from his or her culture nor to self-righteously dismiss other cultures' dances.

Dance as a stand-alone subject is peculiarly important for Zimbabwean learners since, p. "Indigenous Zimbabwean dance equips the learner with a cultural heritage necessary for the continuity of the nation's values, beliefs and practices" (CDTS, 2015, p.1). The Curriculum Framework claims that inter-cultural understanding and tolerance are enhanced "through learning about each other's arts and crafts, music and dances, poems rituals, cuisine and apparel rooted in Indigenous Knowledge Systems" (MoPSE, 2015, p.6). The new dance syllabus, in response to the Curriculum Framework, presents topics, content, activities and resources that could help produce an enlightened graduate who cherishes a Zimbabwean identity. The syllabus also encourages the promotion of linkages with global cultures so that the graduate fits into "an increasingly globalised and competitive environment" (MoPSE, 2015, p.6). The syllabus comes along with equally empowering cross-cutting themes to be studied such as human rights, health, gender, sexuality and ICT.

In Zimbabwe, the schools offering dance vary from the well-equipped elite schools to the under-resourced entrants into the area of dance education. It is crucial to have in place, dance studios with fitted full-length wall-to-wall mirrors, digital video cameras, MP3 players, speakers and other performance aiding learning resources, for an empowering curriculum to be realised. But for the majority of under-resourced rural schools such prerequisites might be absent. Teachers and students in situations such as the latter should not despair but should turn to IKS available in their lived social contexts. Oftentimes, they could use the open grounds and streets to perform dance and also witness communal dance-studded ceremonies. Such experiences could be harnessed during dance lessons through communal and context-based pedagogies, which, in most cases, are epistemologically empowering. For instance, practical dance lessons could be held in the open while resource persons and local talent could be invited to give talks to the learners on the value and dynamics of various dances, hence falling back on indigenous sources of knowledge. An insightful appraisal of a situation can enable a mixture or combination of traditional and modern ways of teaching and learning dance.

The 'appropriate pedagogy' of dance, which is essentially decolonial, needs to be availed if the goal of ethical classroom practice is to be achieved. To this end, the new dance syllabus foregrounds inquiry-based and transformative pedagogies that are also hinged on ICT. More importantly, such an approach resonates with the Freiren critical, collaborative and communicative pedagogy where, according to Amegago (2011, p. xiii), learners are "co-mediators in the learning process". In the teaching of the types and functions of traditional dance the classes should rise above the initial intent to see dance as only politically and culturally liberating. Teachers should strive to raise dance education to a level where it also addresses gender inequalities among other issues. As Styrke (2015, p. 206) argues, dance education should provide participants the opportunity to analyse how choreographers "neutralize, reinforce and/or reduce gendered patterns in their work, and how these issues are addressed." This transformative approach to teaching and learning dance could be pursued in a broadened manner so as to address ethnic, socio-class status, and age and health issues, hence make dance education more pluralistic, yet inclusive. In pursuance of the transformative goal, the syllabus proposes the following activities:

- Listing and categorising indigenous and contemporary dances,
- Demonstrating and performing sequences of dances,
- Presenting repertoires and developing themes or telling a story through dance,
- Discussing performance elements and meanings of dance.

The above activities are transformative and empowering learner-centred pedagogies and at the same time exciting and engaging learners, physically, mentally and emotionally. In addition to the traditionally tested demonstration-reproduction model of teaching, the syllabus embraces the emerging information technology mediated ones that employ resources such as electronic media,

- Video footage,
- Computers,
- CD/DVD/MP4 recorders and players,
- Camcorders,
- JAWS software,
- Smartphones, and
- Graphics

These modern resources would augment the traditional resources such as resource persons, magazines, props, open space, live performance and others. This resonates with what Styrke (2015, p.202) perceives of Sweden after the 2011 reforms on dance education where she notes "an increasing importance of dance as a subject based on artistic as well as science-based teaching." In an ethical pedagogy, teachers and learners should be well aware that dance embodies ideas and practices related to the social and cultural lives of Zimbabweans as well as internal and external political discourses.

The list above is among a host of other challenging ones such as, research on world dance, analysis of choreography and comparisons. The implied significance of the content of the syllabus is its inclusiveness and acknowledgement of other people's knowledge systems and styles which ensures a holistic learning. This is unlike the Eurocentric invalidation and erasure of the dances of the colonised. Thus, in implementing dance education Zimbabwean educators would do well to learn from what Kaola Pratsika meant in the Greek context where she argued that "Dance was important not for its own sake, but for what it could achieve" such as its discursive potential to promote debate on issues of health, heritage, gender and nation, among others (Tsintziloni, 2015, p.277).

CONCLUSION AND RECOMMENDATIONS

This chapter has interrogated the politics of knowledge production to demonstrate the implicatedness of postcolonial curriculum theorising in the quest for ethical pedagogical practice in the Zimbabwean curriculum. The coloniality of power was unpacked in an attempt to illustrate the subtle ways in which the legacies of colonialism permeate the epistemological discourses of the postcolonial curriculum. A critical invocation of IKS as argued herein, does offer a beacon of light on which an ethical curriculum practice that speaks to the lived realities of the Indigenous people could be foregrounded. As argued herein, both History and Dance curricula offer nuanced opportunities for engagement in productive pedagogies that could lead to transformative uncolonial learning which has immense potential for ethical classroom practice. The battle for epistemic freedom from the coloniality of power is not won yet. The imperative on all politically conscious leaders is to be ever vigilant and alive to the matrices of coloniality to seek to perpetuate subservience to Eurocentric epistemologies. For the way curriculum is enacted can radically affect learners as beings located in a world that remains painfully hegemonic. The following areas for possible future research areas can include the politics of the gendered black body and should address the intersection of gender and the black body and the production of worthwhile knowledge. Lastly also on how other school subjects apart from History and Dance are implicated in the politics of knowledge production and their potential towards transformative uncolonial learning.

REFERENCES

Amegago, M. M. K. (2011). An African music and dance curriculum model: performing arts in education. Durham: Carolina Academic Press African World Series.

Asante, K. W. (2000). Dance: Rhythmic forces, ancestral, p. An aesthetic analysis. Asmara, Eriteria: Africa World Press.

Barnes, T. (2007). 'History has to play its role': Constructions of race and reconciliation in secondary school historiography in Zimbabwe, 1980-2002. *Journal of Southern African Studies, 33*(7), 633–651. doi:10.1080/03057070701475740

Bassey, M. O. (2007). What is Africana critical theory or black existential philosophy? *Journal of Black Studies, 37*(6), 914–935. doi:10.1177/0021934705285563

Bhabha, H. (1994). The location of culture. New York, NY: Routledge.

Breidlid, A. (2013). Education, indigenous knowledges, and development in the global south; Contesting knowledges for a sustainable future. London: Routledge.

Bryant, K. D. (2018). Colonial education. In M. S. Shanguhyia & T. Falola (Eds.), The Palgrave handbook of African colonial and postcolonial history (pp. 281-302). New York: Palgrave MacMillan. doi:10.1057/978-1-137-59426-6_10

Chung, F. (2008). Accelerating the expansion of access to secondary education: The 1980 – 1990 experience in Zimbabwe. ADEA – 2008 Biennale on Education in Africa.

Connell, J. (2009). Dance Education: An examination of practitioners' perception in secondary schools and the necessity for teachers skilled in the pedagogy and content of dance. *Research in Dance Education, 10*(2), 115–130. doi:10.1080/14647890903019440

Darder, A. (2016). Foreword: Ruthlessness and the forging of liberatory epistemologies: An arduous journey. In J. M. Paraskeva (Ed.), Curriculum Epistemicide Towards an Itinerant Curriculum Theory. London: Routledge.

Dei, G. J. S. (2014). Indigenizing the curriculum: The case of the African university. In G. Emeagwali & G. S. Dei (Eds.), *African Indigenous Knowledge and the Disciplines* (pp. 165–180). Rotterdam Sense Publishers.

Desai, K., & Sanya, B.N. (2016). Towards decolonial praxis: reconfiguring the human and the curriculum. *Gender and Education, 28*(6), 710–724. . doi:10.1080/09540253.2016.1221893

Elder, C., & Potskin, J. (2018). Interdisciplines, and Indigenous research and methodologies. In C. Lury, R. Fensham, A. Heller-Nicholas, S. Lammes, A. Last, M. Michael, & E. Uprichard (Eds.), Routledge handbook of interdisciplinary research methods (pp. 311-315). New York: Routledge. doi:10.4324/9781315714523-43

Ellsworth, E. (1997). Teaching positions. Difference, pedagogy, and the power of address. New York: Teachers College Press.

Falola, T. (2005). Writing and teaching national history in an era of global history. *Africa Spectrum*, *40*(3), 499–519.

Fanon, F. (2008). Black skin, white masks (C. L. Markmann, Trans.). London: Pluto. (Original publication 1952)

Foucault, M. (1980). Power/Knowledge: Selected interviews and other writings 1972–1977. New York: Pantheon.

Freire, P. (1990). *Pedagogy of the oppressed.* New York: *Continuum.*

Fúnez-Flores, J. I., & Phillion, J. (2019). A Political ontological approach to decolonization of ethnographic research in education. In Rethinking 21st century diversity in teacher preparation, K-12 education, and school policy (pp. 39-54). Cham: Springer. doi:10.1007/978-3-030-02251-8_3

Gittens, A.F. (2012). Black dance and the fight for flight: Sabar and the transformation and cultural significance of dance from West Africa to Black America (1960-2010). *Journal of Black Studies*, *43*(1), 49-71.

Gonye, J. (2013). Mobilizing dance/traumatizing dance: Kongonya and the politics of Zimbabwe. *Dance Research Journal*, *45*(1), 65–79. doi:10.1017/S0149767712000277

Gonye, J. (2015). *Representations of dance in Zimbabwean literature, post 1960* (Unpublished doctoral thesis). University of Venda.

Gonye, J., & Moyo, N. (2015). Traditional African dance education as curriculum reimagination in post-colonial Zimbabwe: A re-think of policy and practice of dance education in the primary schools. *Research in Dance Education*, *16*(3), 259–275. doi:10.1080/14647893.2015.1036020

Gonye, J., & Moyo, N. (2018). African Dance as an Epistemic Insurrection in Postcolonial Zimbabwean Arts Education Curriculum. In The Palgrave handbook of race and the arts in education (pp 157-174). Cham: Palgrave Macmillan. doi:10.1007/978-3-319-65256-6_9

Gottschild, B. D. (2003). The black dancing body, a geography from Coon to Cool. New York: Palgrave. doi:10.1007/978-1-137-03900-2

Greene, M. (1995). Releasing the imagination, essays on education, the arts, and social change. San Francisco, CA: Jossey-Bass.

Gregory, G. J., Luke, A., & Green, B. (2008) Introduction: What counts as knowledge in educational settings, p. Disciplinary knowledge, assessment, and curriculum. Review of Research in Education, 32, p. vii–x. Doi:10.3102/0091732X07311063

Kaya, H., & Seleti, Y. (2013). African indigenous knowledge systems and relevance of higher education in South Africa. *International Education Journal: Comparative Perspectives*, *12*(1), 30–44.

Kincheloe, J. L. (2007). Postformalism and critical ontology- part 1: Difference, indigenous knowledge, and cognition. In J. L. Kincheloe & R. A. Horn (Eds.), The Praeger handbook of education and psychology Vol 4. Westport: Greenwood Publishing Company.

Kincheloe, J. L. (2011). Critical ontology: Visions of selfhood and curriculum. In K. Hayes, S. Steinberg, & K. Tobin (Eds.), Key work in critical pedagogy (pp. 201–217). Rotterdam: Sense Publishers.

Klein, C. (2003). Cold war orientalism; Asia in the Middlebrow imagination, 1945-1961. Berkeley: University of California Press.

Kuwor, S. K. (2003). *Transmission of Anlo-Ewe Dances in Ghana and in Britain: Investigating, reconstructing and disseminating knowledge embodied in the music and dance traditions of Anlo-Ewe people in Ghan.* (Unpublished doctoral thesis). Department of Dance Studies, University of Roehampton.

Leonardo, Z. (2018). Dis-orienting western knowledge coloniality, curriculum and crisis. *The Cambridge Journal of Anthropology*, *36*(2), 7–20. Doi:10.3167/cja.2018.360203

Mabingo, A. (2019). Dancing with the "other": Experiential and reflective learning of African neo-traditional dances through dance education study abroad. International Journal of Education & the Arts, 20(1). doi:10.18113/P8ijea20n2

Malis, M., & Missedja, T.Q. (2019). Schooled for servitude: The education of African children in British colonies, 1910–1990. *Genealogy, 3*(40), 1-12. doi:10.3390/genealogy3030040

Medina, J. (2011). Toward a Foucaultian epistemology of resistance: Counter-memory, epistemic friction, and guerrilla pluralism. *Foucault Studies*, *12*, 9–35. doi:10.22439/fs.v0i12.3335

Mignolo, W. (2009). Epistemic disobedience, independent thought and decolonial freedom. *Theory, Culture & Society*, *26*(7–8), 159–181. doi:10.1177/0263276409349275

Ministry of Primary and Secondary Education: Curriculum Framework for Primary and Secondary Education 2015-2022. (n.d.). Harare: Curriculum Development and Technical Services.

Ministry of Primary and Secondary Education: Dance Syllabus Forms 1 – 4. (n.d.). Harare: Curriculum Development and Technical Services.

Ministry of Primary and Secondary Education: History Syllabus Forms 1 – 4. (n.d.). Harare: Curriculum Development and Technical Services.

MisirHiralall, S. D. (2014). The postcolonial reality of using the term "liturgical" to describe Hindu dance. *Journal of Research on Christian Education*, *23*(2), 154-175. DOI:. doi:10.1080/10656219.2014.899479

Moyana, T. T. (1989). Education, liberation, and the creative act. Harare:. Zimbabwe Publishing House.

Moyo, N. (2014). Nationalist history, nation-state making and secondary school history, p. Curriculum policy in Zimbabwe 1980-2010. *NORDIDACTICA Journal of Humanities and Social Science Education,* 1-24. Available at www.kau.se/nordidactica

Moyo, N., & Modiba, M. (2013). Who does this history curriculum want you to be? p. Representation, school history and curriculum in Zimbabwe. *Yesterday and Today*, *10*, 1–24.

Ndlovu-Gatsheni, S. J. (2013). Coloniality of power in postcolonial Africa: Myths of decolonization. Dakar: CODESRIA Book Series.

Ndlovu-Gatsheni, S. J. (2015). Genealogies of coloniality and implications for Africa's development. *Africa Development. Afrique et Developpement*, *40*(3), 13–40.

Nkomo, M. S. (2011). A postcolonial and anti-colonial reading of 'African' leadership and management in organization studies: tensions, contradictions and possibilities. *Organization*, *18*(3), 365–386. doi:10.1177/1350508411398731

Nyamwaka, E. O., Ondima, P. C., Kemoni, F., & Maangi, E. (2013). The place of music and dance in the reconstruction of African cultural history: A case of the Abagusii of South-Western Kenya. *Research on Humanities and Social Sciences.*, *3*(10), 113–119.

Nyerere, J. (1982). Education in Tanzania (extracts taken from Education for self-reliance). In B. A. Fafunwa & J. U. Aisiku (Eds.), Education in Africa. A comparative survey (pp. 235–254). London: Allen & Unwin.

Nziramasanga, C. T. (1999). Report on the Presidential Commission of Inquiry into Education and Training. Harare: Government Printers.

Odora-Hoppers, C. A., & Makhale-Mahlangu, P. (1998). A Comparative study of the development, integration and protection of indigenous systems in the third world, p. An analytic framework document. Pretoria: Human Sciences Research Council.

Paraskeva, J. M. (2016). Curriculum epistemicide, p. Toward an itinerant curriculum theory. New York: Routledge. doi:10.4324/9781315734781

Richards, K., & Govere, E. (2003). Educational legislation in colonial Zimbabwe (1899-1979). *Journal of Educational Administration and History*, *35*(2), 137-151. DOI: . doi:10.1080/0022062032000119822

Sartre, J. P. (1973). Being and nothingness (H. E. Barnes, Trans.). New York: Washington Square Press.

Shanguhyia, M. S., & Falola, T. (2018). Introduction. In M. S. Shanguhyia & T. Falola (Eds.), The Palgrave handbook of African colonial and postcolonial history. New York: Palgrave MacMillan.

Shizha, E. (2013). Reclaiming our indigenous voices, p. The problem with sub-Saharan African school curriculum. *Journal of Indigenous Social Development*, *2*(1), 1–18.

Silova, I., Millei, Z., & Piattoeva, N. (2017). Interrupting the coloniality of knowledge production in comparative education: Postsocialist and postcolonial dialogues after the cold war. *Comparative Education Review*.

Sithole, T. (2016). Frantz Fanon: Africana existentialist philosopher. *African Identities, 14*(2), 177-190. DOI: 5 doi:10.1080/14725843.2015.111738

Smith, L. T. (1999). Decolonizing methodologies, p. Research and indigenous peoples. London: Zed Books University of Otago Press.

Solórzano, D. G., & Yosso, T. J. (2002). Critical race methodology: Counter-storytelling as an analytical framework for education research. *Qualitative Inquiry*, *8*(1), 23–44. doi:10.1177/107780040200800103

Styrke, B. (2015). Didactics, dance and teacher knowing in an upper secondary school context. *Research in Dance Education*, *16*(3), 201–212. doi:10.1080/14647893.2015.1036021

Trevor-Roper, H. (1963). The rise of Christian Europe. The Listener, 70(1809), 871–875.

Tstintziloni, S. (2015). Koula Prastika and her dance school, p. embracing gender, class and the nation in the formative years of contemporary schooling dance education in Greece. *Research in Dance Education*, *16*(3), 276–290. doi:10.1080/14647893.2015.1036017

Wane, N. (2008). Mapping the field of Indigenous knowledges in anti-colonial discourse: a transformative journey in education. *Race Ethnicity and Education*, *11*(2), 183-197. DOI: . doi:10.1080/13613320600807667

Wane, N. (2011). African indigenous feminist thought: An anti-colonial project. In N. Wane, A. Kempf, & M. Simmons (Eds.), The politics of cultural knowledge (pp. 7-22). Rotterdam: Sense Publishers. doi:10.1007/978-94-6091-481-2_2

Wane, N., & Simmons, M. (2011). Introduction: The politics of cultural knowledge. In N. Wane, A. Kempf, & M. Simmons (Eds.), The politics of cultural knowledge (pp. 1-6). Rotterdam: Sense Publishers.

Widel, A. (2009). British colonial education in Africa: Policy and practice in the era of trusteeship. *History Compass*, *7*(1), 1–21. doi:10.1111/j.1478-0542.2008.00560.x

Wrigely, T. (2003). Schools of hope: A new agenda for school improvement. Stoke-on-Trent: Institute of Education Press.

Yosso, T. J. (2005). Whose culture has capital? A critical race theory discussion of community cultural wealth. *Race, Ethnicity and Education*, *8*(1), 69–91. doi:10.1080/1361332052000341006

KEY TERMS AND DEFINITIONS

Embodied Knowledge: Is the situated ways of knowing that arise primarily from the lived realities of people. It expresses the view that knowing is informed primarily by who we are, and as such, cannot be separated from those who purport to know.

Epistemological Fascism: Refers to willful practices that seek to deny and erase Indigenous Knowledges while promoting Euro-centric views as universal knowledge.

Ethical Practice: Is generally about doing what is considered to be fair and just to all. It can also be applied to determine what is considered right as matter of principle.

Indigenous Knowledge Systems: The sum total of ways of knowing that are ingrained in the everyday cultural practices of a people in a given locality.

Pedagogy: Is primarily about teaching and learning. It also involves content selection as well as the assumptions that inform the selection and choice of methods applied in a classroom context.

Politics of Knowledge: Is about the different and competing views about what ideas are important and should be seen to prevail and hold value over others in their being considered true.

Politics of the Black Body: About how the Western Europeans have conceptualised the African body as different and therefore, inferior and suitable for enslavement and labour through colonisation. This was despite the black body's potential to resist such inferiorisation through inscribing its own history.

Transformative Practices: Are about those ways of thinking and acting that have power to change positively the ways in which people perceive and act in their contexts. It has become a trope for working towards social justice, especially in the classroom context.

Chapter 9
The Ancient Wisdom and Resurgence for Yoga:
A Sociological Perspective

Jah-Xolani Radebe
University of KwaZulu-Natal, South Africa

ABSTRACT

This chapter aims at connecting ancient wisdom and traditions of yoga by looking at the past with fast-paced, cutting edge natural science that constantly re-invents itself with time. It will unpack and explain both wisdom and experience from the orient against science and experiences from the West. The separation of East and West will give us an opportunity to see connections and boundaries of Western lifestyle as a convenient signpost to why yoga is relevant to society today. Furthermore, the chapter seeks to go beyond spirituality, mythology, and present yoga as a perpetual, evolving practice relevant in today's societies to challenge and engage today's science and medical research.

INTRODUCTION

Most of the times general people hold an idea that yoga originated in India and its roots are of Eastern origins, nevertheless some academics and historians suggest that yoga originated in Africa Kemet[1] . Ancient Kemetic people originated Kemetic yoga which was later taken to the shores of the East by ancients Egyptians (Scaramella, 2012). Kemetic yoga focus on the styles that heals and restores body through spinal alignment focuses on relieving stress, improve blood circulation and uses pranayama to smooth the body (Scaramella, 2012). Most of the time, it is challenging to talk

DOI: 10.4018/978-1-7998-1249-4.ch009

about Yoga to the general public and academic audience, especially when it comes to defining what Yoga is and why it is relevant on modern society. One should understand Yoga as a diverse and rich form of exercise. Some of the components of Yoga may include physical, cognitive and psychosocial aspects. Yoga has a long history and evolution and is ever-changing socially and culturally. A range of Yoga styles and applications have developed with slightly different characteristic (Sherman, 2001, Chan, Ng, Chan, Ho & Chan, 2003, Sherman, 2001, Knapik, Steelman, Hoedebecke, Austin, Farina & Lieberman, 2016). The simple broad definition of Yoga is that yoga is a mind-body and soul exercise rooted on Oriental traditions, this includes physical training, traditional medicine and philosophy.

Yoga differs with the normal Western attitude, Eurocentric model of life and philosophy. Third world countries with black population and some of the Western world, looks at yoga as strange practise. Yoga is not glamorous, it is something which is slow and steady, gradual and constant and humble. It requires a transformation of one's mind-set. Yoga takes time. This means, doing yoga is engaging in something for long time to commit on mindful manner (Sherman, 2001, Chan et al, 2003, Sherman, 2001, Knapik, et al, 2016). Practising yoga is based on the science of breathing, therefore breathing supplies every system in our body, each breath deliveries the cells with essential life giving force called oxygen (Sherman, 2001, Knapik, et al, 2016). By focusing our attention on the breathing and flow of breath we help to centre ourselves, strengthens our perception of our surroundings and respond with calmness.

The practice of pranayama[2] is a regulation of life energy by consciously controlling the breath. Research proves that diaphragmatic breathing has an immediate effect on mood and level of a calm person experience and blood circulation (Chan et al, 2003, Sherman, 2001, Knapik, et al, 2016). We should note that shallow chest breathing causes an excess of carbon dioxide in one's bloodstream, which directly affects the acidity level of the blood. Gases imbalance in the bloodstream can leave a person feeling tired, anxious, harried and stressed. When learn to breath and master it, we gain patience with yourself (Sherman, 2001, Chan et al, 2003, Sherman, 2001, Knapik, et al, 2016). Correct breathing is as important as rehabilitating an athlete after an injury. We should always remember that yoga practice is non-judgmental. The essence of mastering the art of yoga is learning to honour oneself and others. Therefore, yoga can be best described as set of living and evolving practise that have been informed by the insight of long lineage of devoted practitioners, moulded and adopted yoga over time to ever changing cultural needs and social landscape (Sherman, 2001, Chan et al, 2003, Sherman, 2001, Knapik, et al, 2016). The diversity and richness of yoga is intertwined in it historical origins.

Today we should focus attention on range components that are the makeup of Yoga, and most of these are therapeutic factors relevant on the psychosocial context of modern society. Perhaps that what makes Yoga unique then any form of physical training? It is multidimensional and holistic exercise seek to integrate individuals psychosocial and philosophical (McLaughlin,Wang, Noone, Liu, Harazduk, Lumpikin, Harmati, Saunders, Dutton & Hakim, 2011, Finkelstein, Brownstein, Lan & Scott, 2007). Yoga effects can are especially important for diseases which are connected with western life nodes for example chronic diseases that connect many systems of the body; this include, the nervous, and immune systems to affect how well the heart function (Howe, Leidel, Krishnan, Weber, Rubenfire & Jackson, 2010, Caldwell, Emery, Harrison & Greeson, 2011, Khalsa, Hickey-Schultz, Cohen, Steiner & Cope, 2012, Ikeda, 2011). Furthermore, yoga seems to be relevant today for social support which includes interactions and community. To be a part of yoga class have proven to be beneficial and therapeutically for range of conditions including social alienation, anti-social, downheartedness, anxiety and life style based ill behaviour. Attending yoga classes teaches practitioner confidence and sense of responsibility as one develops sense of community, with a rich interaction especial from the students and teacher's context which is characterised by support from teacher and peers. Social interaction in context of yoga is great and life changing experience on the passage of self-discover (McLaughlin, et al, 2011, Finkelstein et al, 2007, Howe et al, 2010, Khalsa et al, 2012, Ikeda, 2011). As the student practise with yoga community, yoga creates a practical frame work for practicing living a holistic, and to embrace eastern wisdom that seeks to deal with body mind, and spirit. It can also be deployed as the vehicle to add on spiritual dimension to one's life. Spiritual dimension cannot be negated, on the holistic process of healing prayers and belief is a corner stoner to totally healing (McLaughlin, et al, 2011, Finkelstein et al, 2007, Howe. et al, 2010). Due to emergence of western medical practice the relationship of spiritualty and healing was changed because western medicine practice is not associated with spiritual belief. But today many seek total healing, thru prayer for comfort and solace (McLaughlin, et al, 2011, Howie al, 2010, Caldwell, et al 2011, Khalsa et al, 2012, Ikeda, 2011). Today in the West range of survey reflect that 90 percent Americans populace considers spirituality and God to be significant part of their life (Yeh, McCarthy, Wayne, Stevenson, Wood, Forman, Davis & Phillips, 2008)

The Historical Background of Yoga

The history of yoga is estimated to exist more than five thousand years. Historians and archaeologists prove that the system of yoga was practiced from the ancient times of Kemet-Egypt, Mayan, India and China (Scaravelli, 2012, Radebe, 2018). The focus of Kemet and Oriental wise dome is much more serious about attitude towards our bodies, which we have neglected for many years. The attitude of Kemetic wisdom involve mind, body and spirit, it is our duty today to learn to listen to our body, mind and spirits (Iyenger, 1988, Radebe, 2018, Scaramella, 2012). The Gods and Kings discovered something that make their bodies great, they discover the religion of the body. We have witnessed most Kemet statues depicted the Pharaohs on the dignified posture with spine erect and straight to keep body on its natural sophistication (Scaravelli, 2012, Radebe, 2018).

Kemet statures always shown with upright spines, they did exercise to awaken the body which was good for elevation of mind and spirit. The form of yoga that they were practising was based on raising the energy level and protect one from sicking energy (BKS, 1988, 1996, Scaramella, 2012). Asanas were practised for health promotion and exercises based the postures and movements. For example, Fish Pose Matsyasana, Cat Pose Marjariasana, Down Facing Dog Adho Mukha Svanasana and others reflect extend the tradition of observing nature, mimicking elements of nature and apply them to our health (Iyenger, 1988, Radebe, 2018, Scaramella, 2012) .

The works of Patanjali was remarkable on yoga since he composed the basic text of yoga which was knowledge from older traditions (Scaramella, 2012, BKS, 1988, 1996). In the 1970s, yoga reached new shorelines of the Western region through Hindu teachers such as the late B.K.S Iyenger the author of *Light on Yoga*, and students like Sri T. Krishnamacharya (Scaramella, 2012, BKS, 1988, 1996). On the 70s yoga gained popularity over North America and Europe as some people were eager to learn and find refuge in ancient wisdom from. Yoga has shown dynamic adaptation to changing cultural needs and opportunity to explore alternatives to health. Yoga's expansion to the western hemisphere interface with science and proof-based biomedicine, along with shortcoming of western medicine and search for holistic approach to health, yoga catalysed its resurgence and relevance in the west some (Scully, Kremer, Meade, Graham & Dudgeon, 1998, Ivtzan & Papantoniou, 2014, Scaramella, 2012). Currently, it is reported that in the West more than fifteen million Americans have some exposure to hatha yoga[3], today classes are offered by health clubs and spas, conscious communities, community centres, civic societies, private studio's, organization like Art of Living, and some faith based conscious communities (Radebe, 2018, Wang, Bannuru, Ramel, Kupelnick, Scott, Schmid, 2012).

As a result, in the West yoga is used for physical conditioning and strengthening of the body as it is practiced by many sports personnel and adopted by some governmental institutions. Yoga is taught on hospitals, universities and community centres on the west and only high-class people knows about it on the case of the Third World people (Valvari,1996, Scaramella, 2012, BKS, 1988, 1996). Today yoga is popular because it has blended and collaborated with other practice including Tai-Chi, Qi-gong, meditation, Alexander Technique, mindful exercise and mind-body exercise. Guru like B.K.S Iyenger's compassionate nature to beginners deployed tools[4] which makes yoga asana easier. This kind of yoga involved few asanas, but the gist was to focus more on the perfection of pranayama (Radebe, 2018, Chan et al 2001).

Defining Health and Body Energy

The World Health Organisation defines health with phrases that are forever changing, today there is still no monitoring process for when, and how much, we are health (WHO, 2006, Radebe, 2018). Most people know about health when they are unhealthy, and when there is something wrong or something they do not understand about their bodies and mental state (Radebe, 2018, Wayne et al, 2013, Ivtzan et al, 2014, Macritchie, 1998). The working definition of health in the western medicine is when you are not ill. In the West when someone is ill they ask "What is wrong with this person? And then go about doing whatever it is that is necessary to track down the reason for sickness and cure it, destroy it, remove it or whatever seems relevant according to current level of medical science.

In the East, this whole proposition is contrary, when someone is ill, they ask the simple question "Why this person is not healthy? Because if they were healthy, they would not be ill. The East just ask the same question differently and save hundreds of billion dollars. Eastern definition of health is based on the Energy Systems slight difference and in higher level rather physical system (Herrera, 2015, Kaminoff & Mathews, 2012, Chou, 2005)

ENERGY >>>BLOOD and CELLS>>>ORGANS >>>FUNCTION RELATIONSHIPS>>> THE WHOLE

The ancient energy system functions as a control system for the body, since it lays down the basic framework. The Eastern wisdom of yoga is a best way to boost the body, deep breathing, mixed with strengthen, increases the oxygen in the system and rejuvenates all cells, giving more energy (Kaminoff et al, 2012, Chou, 2005). It is no secret that the western life style causes stress and western medical professionals attempt to reduce stress by prescribing sleeping pills and range of drugs seeking to achieve state of relaxation.

For Eastern wisdom, energy is the ability to do work. Whereas in the West, people repeatedly confuse energy with power and force. However, the awareness of energy is the essence of Eastern knowledge and it infiltrates every walk of life, energy can be accessed in different sources of energy (MacRitchie, 1993, Kent, 2000). The nearest thing to chi or energy in the West is describing how you feel about something. We should note that the human body, energy has a structure or anatomy, such structure is governed by the set of rules and laws (MacLaughlin, et al, 2011, Finkelstein et al, 2007, Wong, 2012). It is the foundation of health, it is what herbs act upon, it is in yoga exercise, and it is what the practice of Tai Chi Chuan effects and develops (Radebe, 2018, Scaramella, 2012, Pegrum, 2001, Lalvani, 2000). The key to this is the understanding of physiology of your energy system. The East wisdom suggested that the state of health is based in our channel system, and that energy can be guided and developed internally by the mind and externally by movement. It is essential to note that, one may come across many unfamiliar concepts that are conventional Western epistemology and there are many words and names for the same thing, which may differ according to the particular source or translation (MacRitchie, 1993, Radebe, 2018 Kent, 2000).

Figure 1.

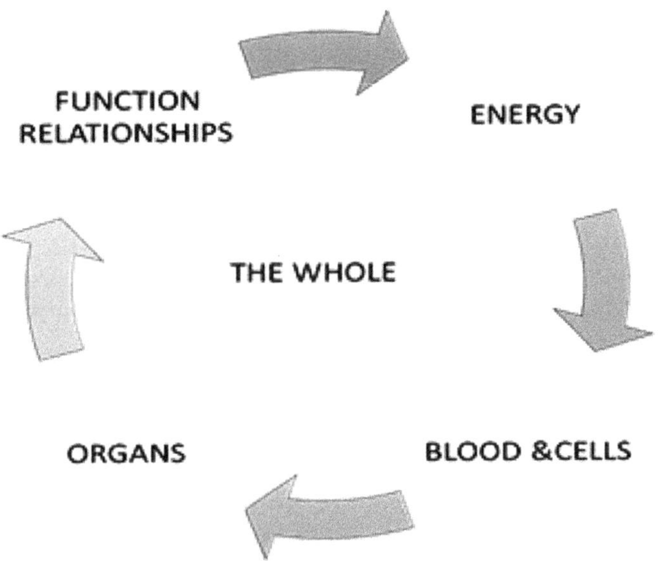

The Eastern wisdom seems relevant today. Eastern energy wisdom is based on resolving symptoms of disease by treating root and cause. The East wisdom focus on health as opposed to management and maintaining of disease as per western epistemology (MacRitchie, 1993, Chan, et al, 2006 Kent, 2000). The basic tenet of the ancient wisdom of is founded upon the belief that human beings hold a unique place in God's creation. As a result, the understanding of the relationship between energy and body systems forms the basic understanding on the links between body, mind and spirit (Ivtzan, et al, 2014).

Humans rely on energy to circulate the blood flow through the heart and the body. It is energy that streams through channels to veins, vessels, capillaries and tissues of major organs. Scientific studies have shown that the movement of energy is essential in blood circulation, along with the physical movement through the heart and vascular system (Herrera, 2015, Nomadlala, Assounga & Hariparshad, 2012, Knapik et al, 2016). It is noted that when energy diminishes through stress, emotions and other negative ways, the flow of life giving force, blood and fluid becomes stagnant. Such stagnation means that the blood flow is decelerating and essential nutrients which the blood and fluids need are slowed also. Good energy, on the other hand, improves blood and fluid circulation and is thus crucial for the healthy functioning of your brain and your whole body (MacRitchie, 1993, Ivtzan, et al, 2014, Kent, 2000, Lomas, 2014, Hefferon et al 2014).

Energy is Everything, One Gets Energy From Different Sources of Energy

- **Food:** Food is one of the main source of energy, right and healthy, balance in food is vital and laws of nature apply "you are what you eat". Food has a powerful effect on the mind and emotions, according to yoga scripture stipulates that consumption can enhance or hinder our energies. It is believed that food is a combination of energies (Radebe, 2018). That is why most yoga practitioners favour vegetarian foods.
- **Rest:** Letting go of the tension in every part of our bodies is an integral part of yoga. Relaxation and rest sometimes it is elusive, even during sleeping hours. Yoga scriptures states that relaxation is not when the body is relaxed but also the mind and spirit are completely at rest (Radebe, 2018). Good and quality rest can provide our bodies with good energy, and can be attainted through training and awareness. Yoga uses physical, mental and spiritual
- **Breath:** Breathing is essence of life. To breathe correct is to breathe with through the nose from diaphragm, Ujjayi Breathing, Deep breathing and through nostrils breathing. Correct breathing is essential to a healthy mind and body. In yoga one uses the breathing technique called pranayama, to balance energies and to focus the mind (Radebe, 2018, Scaramella, 2012). Within our bodies there are seven energy centres called chakras[5] through breathing technique of pranayama one can unlock the blockages so that energy can flow effortlessly from base of the spine to the top of the head crown (Lalvani, 2000)
- **Happy State of mind:** Our mental state of affairs physically and spiritual. It is not easy to always keep mind alert and body to prevent depression. The way to living a happy and less stressful life is to be happy and good vibes will come in (Knapik et al, 2016, Ivtzan, et al, 2014).

It is important to note that the fundamental substances that our bodies need are energy, blood and other fluids. Sources of energy can be obtained from the essence of breathing, eating and drinking. We usually take energy from external sources into ourselves and sustain our function (Scully et al, 1998, Hefferon et al 2014, Herrera, 2015). The quality of the energy our bodies intake is the determinant of the quality of blood, such quality of blood determines the quality of energy; thus the stagnation in one will surely cause stagnation in the other (Kaminoff et al, 2012, Radebe, 2018, Knapik et al, 2016).). Therefore, the essence of food and drinking transformed fluids to nourish and lubricate the organs and tissues. They are food to the cells, having good quality food is essential for one's health and vitality. When one is practicing yoga, one should note that the objective is to deliberately circulate our body fluids through "prana" in order to circulate energy everywhere. By actively directing prana, vitality, and life giving force to the body organs, with our breath through yoga practice allows us to identify areas where energy is blocked, some of the blockages may manifest as emotional discomfort.

So, where does imbalance and disease come from? How is it that people become ill? Why does the chi become dysfunctional? For Eastern wisdom today, there are two basic causes of disease: that is external and internal factors (Ivtzan, et al, 2014, Lomas, 2014). External factors can range from: adverse environmental conditions of heart, cold, wind, dryness and humidity; wrong diet, soiled foods worms and microbes, poising and pollution, trauma and accidents (Ivtzan et al, 2014 Lomas, 2014, Wong, 2012). On the other side, internal conditions can stem from excess or deficient emotions of anger, joy, sympathy, grief or fear- which in turn can generate from emotional environment you are in. They can be caused by inappropriate mental attitude and belief. There are also maladies of the spirit which can cause serious problems. These factors can cause one's energy to become excessive, deficient, stuck, blocked, congested or stagnant, and thereby cause one all manner of problems (Scully et al,1998, Ivtzan et al, 2014). Knowing and understanding this, may help to avoid these factors which cause problems, and can be a major aspect of preserving our energy.

Today, life seems to be unpredictable with modernity, globalization and it globalising forces. Life today is not predictable with promised liberty of the western life. We often here questioning 'Why do bad things happen to good people?' Those who are patients, victims, and family associates often laments after the diagnostic majors of a terminal illness, the search for a reason often ends in vain (Patel, et al, 2012, Kavar, 2012, Wang et al, 2006). The western health care seems to have shortcomings due to the impractical goals and hopes of 'cure for all', redeeming lives and discharging human sufferings.

The inescapable significance of modern medical progressions, prolonged lives also cause lengthy anguishes for people who have to live with durable ill health or be dependent on heavily drugs and medical technologies (Chan & Chow 2006,Scaravelli, 2012, Wang, et al, 2004).). For examples people with diabetes may lose their vision or have their limbs surgically removed. Today we have witness an increasing number of patients that are reliant upon dialysis as a result of chronic renal failure (Madlala et al, 2012, Howard, 2012, Radebe, 2018). People who suffers from cancer have to live with debilities and restrictions due to surgical treatment. Children with inborn syndromes/malformations can live longer but often in a poor health conditions characterised hopelessness. Stroke and heart diseases are also disabling for many as a result, many feel a sense of hopelessness in view of the scale of the disease and inadequacy of western bio medical treatment (Madlala et al, 2012, Little,2016, Arias et al, 2006).

The Relevance of Yoga to This Age

The ultimate goal of the oriental wisdom is to support the healthy production and flow of life force throughout the body. Oriental energy wisdom was based on resolving symptoms of disease by treating both root and cause. The deeper focus was to re-establish health as opposed to management and maintaining of disease. The basic tenet of the ancient wisdom of the orientals was founded upon the belief that human beings hold a unique place in God's creation (Scully et al, 1998, Ivtzan et al, 2014 Lomas, 2014, Hefferon et al 2014). Therefore, the understanding of the relationship between energy and body systems forms the basic understanding of the links between body, mind and spirit. These links are a vital in the maintenance of health.

Today, many people turn to do yoga as the way to manage their stressful life, practising yoga as alternative remedy to stress is a good reason to why yoga is relevant today. It is proven by many medical scientist that yoga can remedy or reduce stress by calming the mind, and create peace and relaxation (Jadhav et al, 2009, Scully et al, 1998, Lomas, 2014, Hefferon et al 2014). Yoga has many reasons to why it is relevant today, its benefits factors are synergistic and that's what make yoga a peculiar exercise. Yoga should be taken as the multicomponent exercise which seeks to effect physical, psychological, social, and philosophical level. It is multidimensional effects vital to curtail and cure chronic disease that impacts many systems; respiratory immune system, endocrine system and heart function (Wang, 2012, Scully et al, 1998, Hefferon et al 2014).

Hatha yoga and pranayama is relevant today and can help today that often places us under great stress and strife. Most people constantly feel exhausted and stressed from work, with no time for friends and loved ones (Scully et al, 1998, Lomas, 2014, Hefferon et al 2014). Our bodies often judged us because of the unnatural postures that we adopt: longer hours from driving and odd poses while working on the computer machine. Our body health system becomes poor, as we are subjected to pressure due to unhealthy life style habits resulting in bodily disorders (Scully et al, 1998, Jadhav et al, 2009, Wang 2012). The modern food leaves us longing deadly but nice tasting foods. Most people constantly feel gastral pains if not mental discomfort without the use of drugs in the form of coffee, alcohol, medication and other substance abuse.

Yoga seeks to maintain right health; "the right state of mind, free from social and personal confusion, the right state of stable emotions, being grounded, having a desirable state of being, feeling well integrated under universe" (Radebe, 2018:3, Jadhav et al, 2009). Yoga practitioner seeks clear-mind and mindfulness, happiness, satisfaction and thrives to attain higher self, and spiritual development. It can be probably possible be realized, or improve through all the above with nothing else but the practice of yoga and pranayama (Wang, 2012, Scully et al, 1998, Ivtzan et al, 2014)

Today medical research, doctors and other sciences confirm that practicing yoga cures illness, prevents diseases, strengthens the body and heightens the response of the immune system, prevents premature aging and promotes a prolong life span (Ivtzan et al, 2014 Lomas, 2014, Hefferon et al 2014). By doing yoga asana, one can undo the damage rather than just turn a blind eye on symptoms caused by an undesirable lifestyle. Yoga can help us find the root of causes of our ailments in life, and it's guaranteed to make permanent changes (Jadhav et al, 2009, Wong, 2012). It is the science or art that goes beyond just postures we see being performed in yoga classes. Yoga cure illnesses and provide a ground-breaking health care and awareness of health. It can decrease stress and reduces stress charges (Scully et al, 1998, Ivtzan et al, 2014 Lomas, 2014, Hefferon et al 2014).

Society suffer from cardiovascular diseases and high blood pressure, headaches and chest pains, diabetes and obesity which are all connected to stress and may be very expensive when it comes to medical treatment (Scully et al,1998, Ivtzan et al, 2014, Hefferon et al 2014). This could be much pursued by adopting preventative means that will always keep you in sound and good health. For example, by changing meat heavy western diet to prevent obesity which results to leading an unhealthy risky lifestyle (Wong 2012, Scully et al, Lomas, 2014, Hefferon et al 2014). Once more yoga can help us to develop a strong foundation of health. Today, obsession on health and fitness, health clubs, excises and sports and other alternative ways to develop strong health. The evidence based research proves that yoga is a safe way of to rejuvenate, heal and restores your energy from nature and guide energy through every cell of the body from your crown of your head, moving down through the chest, shoulder and spine, and all the internal organs though legs and feet (Scully et al, 1998, Ivtzan et al, 2014 Lomas, 2014, Hefferon et al 2014, Wong, 2012). Yoga practice can cultivate enhanced levels of satisfaction in life and subjective happiness supports studies indicating that yoga-based interventions can promote psychosocial wellbeing (Ivtzan et al, 2014 Jadhav et al, 2009, Wang 2012).

Today yoga is popular because people are seeking healthier and lifelong solution to the stresses that are connected to the modern living conditions. Yoga is not only about body, but one can gain strength and clarity of mind. Yoga is lengthening practices that can carry on throughout your life without regard to age. Everyone can benefit from this ancient wisdom of yoga from: student who are stressed and working hard for desirable outcomes on exams, housewives with lots on their hands as looking children and home chores, working people who are trying to maintain a career, executive who need to stay on top stressful work setting, and senior citizens facing challenges with ageing and retirement. Yoga is a way of life rather than religion, and its does not interferes with anyone's belief system. Yoga philosophy is based on the tenant that all challenges can be solved by building inner strength and character. Yoga teachings can transform people's lives by teaching a new way of thinking and viewing the world surrounds us (Ivtzan et al, 2014 Lomas, 2014, Hefferon et al 2014).

Social Impacts of Yoga on Community

Life today is considered to be stressful, people are too busy with working to achieve life demands survival. The competitive environment and technological evolvement results to stress (Mthembu, 2016, Straume, et al, 2012, MacLaughlin, et al, 2011. Industrial society work style is linked with unhealthy eating pattern and family life is a completely reason why people are stressed and feel the pressure in life (Mthembu, 2016, Nicole De Wet, 2017 Straume, et al, 2012). The lifestyle brought stress to the individuals which leads to diseases, consequently, practicing yoga is relevant since it provides holistic health solution physical, psychological, social and spiritual (Nicole De Wet, 2017). Yoga benefits to participants can:

- **Cure Disorders:** This is significant to everyone; yoga can provide a revolutionary health care and awareness health, to everyone who practices yoga.
- **Reduces Stress and Reduces Stress Cost:** Most people suffer cardiovascular diseases and high blood pressure, headache and chest pains, diabetes and obesity can connect to stress.
- **Prevention of Diseases:** this could be the much pursued after preventative means that will always keep you in sound and good health. Example obesity and meat heavy western life style, unhealthy risky life style.

- **Strengthening your constitution:** yoga can help to develop a strong foundation of health. Today we have witness the obsession health and fitness, health clubs, excises and sports.
- **Avoiding Premature Aging:** We all subjected to ages, it one of nature's rule, but then again no one wants to age any faster than they can possible avoid. Asana and pranayama helps to avoid premature aging.
- **Rehabilitate People Living with Drug Addictions:** Yoga offers mind power and detoxification of the body through traditional medicine and physical exercises to help fighting addictions.
- **Improve Your Sexual Life:** Sexual energy is vital force we have, it is energies of putting life and we establish deep connection between individuals. Sexual energy is a source of insecurity and dissatisfaction for many today.

Yoga is gaining popularity today because people are seeking healthier and more long-lasting solutions to the stresses coursed by western modern life and culture. Yoga is not just exercise, it is discipline that seeks to achieve a well-toned body but also gain a strength and clarity of body, mind and soul (Jadhav et al, 2009, Straume et al, 2012). Yoga practice is timeless and regardless of age, anyone can benefit from it. From the student who are working towards their exams, housewives who are assigned with duties of child minding and home chores, mothers who are working hard to maintain high managerial jobs and senior citizen who are challenged by retirement and facing aging stages. It is important and to note that yoga is a knowledge and philosophy of life rather a religion, as a result it does not interfere with anyone's life, belief and religious beliefs (Straume et al, 2012, Wong, 2012). Yoga philosophy is based on believing that all challenges can be solved by developing inner strength and character. It is truth that everyone's life is characterised by up and downs, and it is hard sometimes to overcome. Today people can learn that yoga transforms people's lives simple by teaching them a new way of thinking and view of the world, giving people strength, courage and hope in an increasingly chaotic age (Ivtzan, et al, 2014, Wong, 2012, Jadhav et al, 2009).

The way people react to stressful situation differs widely and any factors influences the way one act. Much depends on environmental factor, genetic differences, age, sex, family circumstances, food and occupation. Usually, any lifestyle changes will always result to stress if a person cannot adapt to the changes. Yoga therapy is important today with it unique contribution that is directly affects the brain, especially the psychic centre from where psychosomatic stress disorder are initiated (Lalvani, 1999, 2000, Pergrum, 2001, Straume et al, 2012, Radebe, 2018). It is for this reason that yoga helps to manage stress related ailments. Today we are living in an ever-changing society with pressure put upon individuals to succeed. This is the reason why certain disorders in developed and developing countries disorders is increasing disorder are at an alarming rate (Ivtzan, et al, 2014, Straume et al, 2012). However, yoga can play an important role in helping society to maintain harmony and balance; people who practice regular are more likely to be in tune with themselves and their surroundings and lead much happier life (MacLaughlin, et al, 2011, Finkelstein et al, 2007). People who suffers from stress, on the other hand, display stress on their posture face, breathing patterns, and facial expressions (Wong, 2012, Straume, et al, 2012). Maintaining good health, mental health by eating a balance diet is a vital part of yoga. Over indulging in food, alcohol and smocking burdens body with a heavy overload and forces body system to work twice as hard to cope (Radebe, 2018, Lalvani, 2000). People who suffers from stress can complain about fatigue, depression and generally feeling unwell. They usually lead inactive lives with little exercise, and they seem to have no joy in anything they do. Therefore, the lack of energy affect mental health as it erodes self-esteem. In this chaotic modern world, ancient wisdom and philosophy can restore harmony, balance, and bring peace back to the soul (Straume et al, 2012, Wong, 2012, Radebe, 2018)

Revitalization and Necessity for Ancient Wisdom of Yoga

One of the short comings of the Western psychosocial care model, is it focus on single component mediation seeking to change a speciðc behaviour and alter cognitive appraisal or promote emotional expression. Today, alternative to healing connect traditions seems to be relevant since it looks at the health problems with the holistic view (Ivtzan, et al, 2014). For Eastern wisdom a person is exclusive subjected subtleties of multifaceted personal-relations, environment seen and unseen, physical and philosophical-moral interactions with self, other and with the universe (Radebe, 2018, Ivtzan, et al, 2014, Knapik et al, 2016). Instead, western intervention model concerns only with the focus on antagonizing the ailment or dealing with that single problem. Eastern model suggest that problems and all ailments are healthier dispensed using the strength and combined effect.

Today most people in the West and Third World countries are looking upon the Eastern sense of healing through a holistic intervention charter (Radebe, 2018). Eastern spirituality entails the faith of inner strength of a person, belief in the naturally compassionate and to be considerate human nature, to come clear about dynamic state of perpetual disparity in search of stability and harmony (Chan et al, 2006, Ivtzan, et al, 2014, Knapik et al, 2016).) Society today is confounded with pandemonium and irregularity of life, the sufferers form modern western life style are searching for meaning and reasons for living. The premise of the eastern wisdom is that we should acknowledge that our suffering can enable to reconsider our life goals, both yoga scriptures and Bhudda scriptures consolidate that pain and grief can help one to reñect upon what is important in life. (Chan et al, 2006, Birdee et al, 2009, The RZA, 2009, Howard, 2012). As a result, ones suffering can leading eventually to rearrangement of priorities.

For Eastern holistic perspective seclusion, holy rites and introspection, is a cornerstone for individuals to be alleviated to higher level of integrity and experience spiritual redirection and growth. In recent times, there has been an increasing awareness of the importance of spirituality in health professional services, Spirituality and holistic interventions have been combined in psychotherapy. (Chan et al, 2006, Chan, et al.2006, The RZA, 2009, Sherman 2001, Miller 1999). For example, mindfulness cognitive behavioural therapy, which harnesses the spiritual peace brought about by yoga through meditation, is an innovative alternative to conventional psychosomatic therapies for major depression (Radebe, 2018, Little, 2016, Hayes et al. 2004). The Eastern body–mind–spirit intervention approach in clinical practice, attempts to enhance the spiritual strengths of patients and under-privileged populations (Chan 2001, Chan et al, 2006, Chan et al, 2003).

The Eastern model seems convincingly holistic, the body–mind–spirit holistic model of eastern philosophies and tradition address physical, mental and spiritual needs of an individual. Eastern holistic approach to wellbeing covers the ultimate and general components of health. Eastern emphasises on the need for spirituality in health promotion of individual's resilience and health in the face of harsh conditions (The RZA, 2009, Chan, 2001, Scaravalla, 2012). The teachings of yoga always based on changing harsh conditions into progress and divine existence. It is on this sense that yoga is an intersectionality of the spirit, mind and body. Today health care systems in many countries are seriously under risk by bureaucratization, and eastern body mind and spirit model will play a signiðcant part in stimulating a holistic health system which can be in congruence with the structure as a whole (Madlala et al, 2012, Chan, et al, 2006, Lalvani, 1999).

The clinical essence of yoga and yoga philosophy advocates a life of non-attachment, and it stipulate to live and found on fully present and living in the moment (The RZA, 2009, Pegrum, 2016). The East wisdom seems to intrude on the western opinion of hope and future-focus. From Zen philosophy to yoga philosophy they all concern about living in the present moment. The life on non-attachment stipulates that, were so worried by the reminiscence of the past or fretfulness about the future which is not yet came (Sherman, 2001, Chan et al, 2003, Sherman, 2001). Most people incompetent to fully live and embrace now or present moment, they are frequently lost in the present moments (Chan et al, 2001 Ivtzan, et al, 2014). The Eastern masters and scriptures simple advocate to concentrate on what we are doing on the moment right now; If you eating your food one should be mindful and be present on the moment of mastication and swallowing during the meals, when you walking focus on the awareness of the muscles while walking. The reason behind such deep thought is to do one thing at a time; and the promise is that the quality of our lives may be different on the positive aspect (Sherman, 2001, Chan et al, 2003, Sherman, 2001, Knapik, et al, 2016).

Need and Importance of Yoga Education

Yoga education can develop our ability to be peaceful at all times in all our actions connected to physical and mental health. Yoga education can play a vital role in the children's development and personality of a child. It is very essential to understand the importance and need of yoga education (Venkatesh, 2005, Singh, 2016, Surdarshan et al, 2003). The important aspect is always our physical wellbeing as individuals. Therefore, yoga education can provides physical growth and personal development, as we learn to develop endurance and strength when doing normal chores of life as well as to meet the demands of the pressure of life, without felling of stress and develop a proper and sound functioning of organs and entire body system (Caldwell, et al 2011, Khalsa et al, 2012, Ikeda, 2011).Yoga education can balance medical technology with a holistic approach of health which seek to address the complications of the mind and spirit. Anyone can learn to control mind, desires, and reaction towards stress. Yoga is an intellectual skill for calming down the mind, yoga education can improve motor qualities such as stamina, power, quickness, flexibility and power (Venkatesh, 2005, Singh, 2016).These qualities are essential for healthy and productive life in society. Performing variety of asanas, can be beneficial to novices as they can learn the essence of supportive and team work. Education and yoga can develop skills which can concoct students for a vocation (Venkatesh, 2005, Singh, 2016).

It is believed that yoga can impact our health—both physical and mental. In our contemporary western culture, we should consider traditional methods and techniques of working with body, mind and spirit. There has been a need for new version or alternatives to the health over last few years, natural medicine and holistic health is suggested to complement western biomedical methods. There are grass root movements that appears around the world, they dubbed as "The New Age" movement (Wang et al, 2010, Merrill et al, 2011, Radebe, 2018). These includes environmental, food and security movement, land movement, hippie culture, Rastafarians, grassroots movements, community activist, student activism, street shouters, conscious communities, civic society, eco political movements, social justice movement, feminist and gander mainstreaming movements, human rights activism that advocate for alternative and back to nature movement (Palumbu, 2010, Merrill et al, 2011, Berry, 2010, Radebe, 2018).

In Kenya Yoga program increased to Kenyans for they have achieved well-being through the use of yoga and their practice produced biopsychosocial benefits, and prosocial, community engagement activities (Giambrone, et al, 2018, Knapik et al,

2016). A sense of community and connection was achieved through training local people to teach yoga. Yoga can help people and families in manage matters while they trying to adjust to potential dramatic lifestyle changes (Wang et al, 2010, Berry, 2010, Clark et al, 2011). A holistic therapeutic effort which is multi-disciplinary is required to minimise the emotional impairment that modernity created. People with chronical illness feel, unemployed, and those suffers from stress feel shock, betrayed, angered when learning that they are in life failure (Madlala, et al,2012, Merrill et al, 2011, Berry, 2010, Clark et al, 2011). Family members often feel the same way, and experience feeling of guilty and shame and self-blame. The yoga role is to assist people to learn with patient in redefining their life goals in the limitation enacted by western life and culture (Madlala et al 2012, Wang et al, 2010, Berry, 2010, Clark et al, 2011). Yoga education can help one to elevate intra-personal sub-system that include emotional and psychological factors as well as interaction with social and physical environment. The goal of yoga is change cognitive thinking and understanding modern life, in doing so this can help a society to feel less depressed about changes in life and enables them to make a behaviour changes in life (Wang et al, 2004, MacLaughlin, et al, 2011, Ikeda, 2011).

CONCLUSION

We should understand yoga as scientific discipline that get rid of or exterminating stress and its developments. Yoga is a constant process and the concept has go through reflective revolution as the change in the world pose a greater challenge to our communities (MacLaughlin, et al, 2011, Finkelstein et al, 2007, Howe, et al, 2010). Yoga training programmes can improve the behaviour and cognitive development. Yoga offers the modest, safe and practical techniques for all people to quickly reduce tension and stress at any time and place. Yoga and meditation, is a pathway to self-knowledge. It takes mindfulness from the outside to inside and passes a realization that the only true gratification grew from knowing that peace is with oneself (Caldwell, et al 2011, Khalsa et al, 2012, Ikeda, 2011). Yoga reduces the effects of stress and persuades a feeling of tranquil and serenity, combats depression and anxiety, counteracts helplessness and weakness, and thus increases self-esteem and internalized locus of control.

The paradox of the fast-pace way of life and culture of the Western culture seems to bound to accept and adopt Eastern wisdom practices such as yoga (MacLaughlin, et al, 2011, Finkelstein et al, 2007). Yoga seems thrives to serve as the crucial point of a controversial debate about yoga in the West, when compared against the classical Eastern body energy system and ancient wisdom (Khalsa et al, 2012, Ikeda, 2011). The West has a need for spiritual and physical intentions to commit to yoga, this chapter offered a unique perspective on yoga. It explained that yoga asana practitioners are gradual evolving to their spiritual intentions, nurtured through sustained practice of asana. The yoga can cultivate states of divine existence. On addition importantly, this chapter emphasises the essential dimension which was overlooked on both side's Western life style and East health system. Lastly, it explained how yoga can be used for educational purposes and for human development.

REFERENCES

Arias, A. J., Steinberg, K., Banga, A., & Trestman, R. L. (2006). Systematic review of the efficacy of meditation techniques as treatments for medical illness. *Journal of Alternative & Complementary Medicine, 12*(8), 817-832.

Berry, L. L. (2010). What is the Hard Return on Employee Wellness Programs? *Harvard Business Review*. PMID:21188899

Brazier, A., Mulkins, A., Sherman, A. C., & Thomas, P. G. (2001). *Research on Faith and Health: New Approach to Old Questions Psychological Perspectives*. Guilford Press.

Caldwell, K., Emery, L., Harrison, M., & Greeson, J. (2011). Changing in Mindfulness, Wellbeing, and Sleep Quality in College Students Through Taijiquan Course: A Cohort Control Study. *Journal of Alternative and Complementary Medicine (New York, N.Y.), 17*(10), 931–938. doi:10.1089/acm.2010.0645 PMID:21999153

Chan, C., Ng, E., Chan, C., Ho, P. C., & Chan, T. (2006). Effectiveness of psychosocial group intervention for reducing anxiety in women undergoing in vitro fertilization: A randomized controlled study. *Fertility and Sterility, 85*(2), 339–346. doi:10.1016/j.fertnstert.2005.07.1310 PMID:16595209

Chan, C. L. W., Fan, F. W., & Gong, R. Y. (2003). *The Body-Mind-Spirit Integrative Health Approach: Group Counselling Theory and Application*. Ethnic Publishing. (in Chinese)

Chan, C. L. W., Ho, P. S. Y., & Chow, E. (2001). A body-mind-spirit model in health: An eastern approach. [PubMed]. *Social Work in Health Care, 34*(3-4), 261–282. doi:10.1300/J010v34n03_02

Chou, L. (2005). *The Martial Artist's Book of Yoga: Improve Flexibility. Balance and Strength for Higher Kicks, Faster Strikes, Smoother Throws, Safer Falls, and Stronger Stances*. Academic Press.

Clark, M.M., Warren, B.A., Hogen, M.D., Johnson, B.D., Jenkins, S.M., Wernerburg, M.S., & Olse, M.D. (2011). Stress Level, Health Behaviour, and Quality of Life in Employee Joining a Wellness Centre. *American Journal of Health Promotion, 26*(1), 21-25.

De Wet, N. (2017). Family formation and economic insecurity among youth in South Africa. *Development Southern Africa, 34*(6), 694–703. doi:10.1080/0376835X.2017.1310028

Finkelstein, C., Brownstein, A., Lan, Y. L., & Scott, C. (2007). Anxiety and Stress Reduction in Medical Education: An Intervention. *Medical Education, 41*(3), 258–264. doi:10.1111/j.1365-2929.2007.02685.x PMID:17316210

Giambrone, C.A., Cook-Cottone, C.P., & Klein, J.E. (2018). *The Africa Yoga Project and Well-Being: A Concept Map of Students' Perceptions*. doi:10.1111/aphw.12124

Herrera, L. L. (2015). *Meridian Systems Yoga: An Innovative & Accessible Method That Combines Yoga & Traditional Chinese*. Medicine Create Space Independent Publishing Platform.

Howe, M., Leidel, A., Krishnan, S. M., Weber, A., Rubenfire, M., & Jackson, A. E. (2010). Patient Related Diet and Excise Counselling: Do Providers' Own Lifestyle Habits Matters? *Preventive Cardiology, 13*(4), 180–185. doi:10.1111/j.1751-7141.2010.00079.x PMID:20860642

Ikeda, A. (2011). Optimism in Relation to Inflammation and Endtheial Dysfunction on Older Men: The VA Normative Aging Study. *Psychosomatic Medicine, 73*, 664–671. doi:10.1097/PSY.0b013e3182312497 PMID:21949417

Ivtzan, I., Lomas, T., & Hefferon, K. (2014). *Applied Positive Psychology: Integrated Positive Practice*. Sage Publications.

Ivtzan, I., & Papantoniou, A. (2014). Yoga meets positive psychology: Examining the integration of hedonic (gratitude) and eudaimonic (meaning) wellbeing in relation to the extent of yoga practice. *Journal of Bodywork and Movement Therapies, 18*(2), 183–189. doi:10.1016/j.jbmt.2013.11.005 PMID:24725784

Jadhav, S. G. U., & Havalappanavar, N. B. (2009). Effects of yoga interventions on anxiety and subjective well-being. *Journal of the Indian Academy of Applied Psychology, 35*, 27-31.

James, M. (1993). *Health Essentials Chi Kung Cultivate Personal Energy*. Element.

Kaminoff, L., & Mathews, A. (2012). *Yoga Anatomy. Your illustrated guide to postures, movements, and breathing techniques*. Human Kinetics.

Kent, H. (2000). *Yoga An Introduction Guide to Optimum Health for Mind, Body and Spirit*. Academic Press.

Khalsa, S. B., Hickey-Schultz, L., Cohen, D., Steiner, N., & Cope, S. (2012). Evaluation of the Mental Health Benefit of Yoga in Secondary School: A Preliminary Randomized Controlled Trail. *The Journal of Behavioral Health Services & Research, 39*(1), 80–90. doi:10.100711414-011-9249-8 PMID:21647811

Killingsworth, M. A., & Gilbert, D. T. (2010). A Wondering Mind is un Happy Mind. *Science, 330*(6006), 932. doi:10.1126cience.1192439 PMID:21071660

Knapik, J. J., Steelman, R. A., Hoedebecke, S. S., Austin, K. G., Farina, E. K., & Lieberman, H. R. (2016). Prevalence of dietary supplement use by athletes: Systematic review and meta-analysis. *Sports Medicine (Auckland, N.Z.), 46*(1), 103–123. doi:10.100740279-015-0387-7 PMID:26442916

Lalvani, V. (1999). *Yoga for Stress: Relax and re-energize the natural yoga way*. Octopus Publishing Group.

Lalvani, V. (2000). *Complete book of Yoga. The total yoga workout for mind body and spirit*. Octopus Publishing Group.

McLaughlin, B. W., Wang, D., Noone, A. M., Liu, N., Harazduk, N., Lumpikin, M., Harmati, A., & Saunders, P. (2011). Stress Biomarkers in Medical Student Participating in a Mind Body Medicine Skills Program. *Evidence-Based Complementary and Alternative Medicine, 2011*, 1–8. doi:10.1093/ecam/neq039

Miller, W. R. (1999). *Integrating Spirituality into Treatment: Resources for Practitioners*. American Psychological Association. doi:10.1037/10327-000

Mthembu, N. (2016). Youth challenges in the transition to adulthood in a South African township. Youth Voice Journal.

Nomadlala, N, Assounga, A., & Hariparshad, S. (2012). *Kidney and Diseases in an African Setting*. Reach Publishers.

Palumbu, M.V. (2010). Tai Chi for Older Nurses: A Workplace Wellness Pilot Study. *Applied Nursing Research, 25*(1), 54-59.

Patel, N. K., Newstead, A. H., & Ferrer, R. L. (2012). The effects of yoga on physical functioning and health related quality of life in older adults: A systematic review and meta-analysis. *Journal of Alternative and Complementary Medicine (New York, N.Y.), 18*(10), 902–917. doi:10.1089/acm.2011.0473 PMID:22909385

Piet, J., & Hougaard, E. (2011). *The Effect of Mindfulness-based Cognitive Therapy for Preventation of Relapse in Recurrent Major Depressive Disorder: A Systematic Review and Meta-analysis*. Clinical.

Radebe, X. (2018). New Yoga 4 New Age: Activate Your Personal Energy. Authors House.

Saunders, P. A., Tractenberg, R. E., & Chertji, R. (2007). Promoting Self-awareness and Reflection through an Experiential Mind-Body Skills Course For First Year Medical Students. *Medical Teacher, 29*(8), 778–784. doi:10.1080/01421590701509647 PMID:17852720

Scully, D., Kremer, J., Meade, M. M., Graham, R., & Dudgeon, K. (1998). Physical exercise and psychological well-being: A critical review. *British Journal of Sports Medicine, 32*(2), 111–120. doi:10.1136/bjsm.32.2.111 PMID:9631216

Singh, S. (2016). Implementing Yoga Education in schools for holistic development of students. *International Journal of Advanced Educational Research*. www.newresearchjournal.com/educational

Straume, L. V., & Vitterso, J. (2012). Happiness, inspiration and the fully functioning person: Separating hedonic and eudaimonic well-being in the workplace. *Journal of Positive Psychology, 7*, 387-398.

Surdarshan, P. V., & Balakrishanaiah, S. (2003). *Physical Education Curriculum at Lower Primary Stage- An Evaluation. Research Project*. Regional Institute of Education.

Venkatesh, M. (2005). *The effect of yoga on the personality development of students* (Ph.D. thesis). Swami Vivekananda Yoga Anusandhana Samsthana.

Verhoef, M. (2006). Evaluating a yogic breathing and meditation intervention for individuals living with HIV/AIDS. *American Journal of Health Promotion, 20*(3), 192–195. doi:10.4278/0890-1171-20.3.192 PMID:16422138

Verhoef, M. J., Mulkins, A., & Boon, H. (2005). Integrative health care: How can we determine whether patients benefit? *Journal of Alternative and Complementary Medicine (New York, N.Y.), 11*(supplement 1), s-57–s-65. doi:10.1089/acm.2005.11.s-57 PMID:16332188

Wang, C., Bannuru, R., Ramel, J., Kupelnick, B., Scott, T., & Schmid, C. H. (2010). *Tai Chi on Psychology Well-being: systematic review and meta-analysis.* ISCMR.

Wang, C., Collet, J. P., & Lau, J. (2004). The effect of Tai Chi on health outcomes in patients with chronic conditions: A systematic review. *Archives of Internal Medicine*, *164*(5), 493–501. doi:10.1001/archinte.164.5.493 PMID:15006825

Wong, P. T. P. (2012). *The human quest for meaning: Theories, research, and applications*. Routledge.

Yeh, G. Y., McCarthy, E. P., Wayne, P. M., Stevenson, L. W., Wood, M. J., Forman, D., Davis, R. S., & Phillips, R. S. (2010). Tai Chi Exercise in Patients with Heart Failure: A Randomized Clinical Trail'. *Archives of Internal Medicine*, *171*(8), 750–757. PMID:21518942

KEY TERMS AND DEFINITIONS

Asana: Are body postures, asanas are called yoga poses or yoga postures in English. There many types of poses or positions, stand-up, twisting, and balancing poses.

Body Energy System: Are 7 centres of energy called Chakras that run from the base of the spine to the top of the head. Each energy centre in the body is believed to create health and energy when open. Body gets its energy from many sources including food, breathing, resting or meditation.

Mindfulness Excises: Are type excises that gives you a chance of meditation in which you focus one to be in the moment, without any judgment. Mindfulness excises includes breathing methods and other practices to lower the body stress and calming the mind to help reducing stress.

Yoga: The term yoga is a Sanskrit word meaning "yuk" meaning union. The union of both physical exercises (asana), mental meditation, and breathing techniques (pranayama) to support the muscles and discharge anxiety.

Yoga Education: Yoga education is the educational appendage to school and university knowledge. This education to give students physically and mentally for the integration of their physical, mental and spiritual faculties so that the students can be integrated to society with a sober mind of yoga and meditation.

ENDNOTES

1 The term Kemet literally means 'Black Land' this name is associated with the colour of the land of Kemet which was black and known to be fertile and rich due to Nile River.

2 Pranayama is the conscious awareness of breath: the life force that both energizes and relaxes the body. The term is derived from the Sanskrit, prana, meaning "life force," and ayama, meaning "extension." Pranayama is a gist of yoga. The controlled breathing enables both the rhythm of performing yoga poses and relaxing the mind for meditation (Radebe, 2018:138-139).

3 Hatha Yoga simply refers to the practice of physical yoga postures, meaning your Ashtanga, vinyasa, Iyengar and Power Yoga classes are all Hatha Yoga. The word "hatha" can be translated two ways: as "willful" or "forceful," or the yoga of activity, and as "sun" (ha) and "moon" (tha), the yoga of balance. Hatha practices are designed to align and calm your body, mind, and spirit in preparation for meditation.

4 Yoga tools like bricks, blankets and anything that can assist beginners to easy up asana.

5 Chakras literally means "wheels" these are energy centres situated along with subtle body.

Compilation of References

Abbatt, F., & McMahon, R. (1993). *Teaching Health Care Workers*. Macmillan.

Acharya, A. (2018). Introduction: Rethinking Agency and Change in Global Order. In *Constructing Global Order: Agency and Change in World Politics* (pp. 1–32). Cambridge University Press. doi:10.1017/9781316756768.002

Adeyemi, M. B., & Adeyinka, A. A. (2002). Some key issues in African traditional education. *McGill Journal of Education*.

Aiken, L. R. (2004). *Morality and ethics in theory and practice*. Charles Thomas.

Akoojee, S., & Nkomo, M. (2007). Access and quality in South African higher education: the twin challenges of transformation. *SAJHE, 21*(3), 385-399.

Alatas, S. H. (2004). The captive mind and creative Development. In P. N. Mukherji & C. Sengupta (Eds.), *Indi-geneity and Universality in Social Sciences: A South African Response* (pp. 24–31). Sage.

Allen, R. E. (Ed.). (1990). *The Concise Oxford Dictionary of Current English* (1st ed.). Clarendon Press.

Amegago, M. M. K. (2011). An African music and dance curriculum model: performing arts in education. Durham: Carolina Academic Press African World Series.

Ani, M. (1994). *Yurugu: An African-centered Critique of European Cultural Thought and Behavior*. African Word Press.

Anyansi-Archibong, C. B. (2015). Challenges in Ethics Education: A Process and Content Analysis. In C. B. Anyansi-Archibong (Ed.), *Contemporary Issues Surrounding Ethical Research Methods and Practice* (pp. 250–273). IGI Global. doi:10.4018/978-1-4666-8562-8.ch011

Arias, A. J., Steinberg, K., Banga, A., & Trestman, R. L. (2006). Systematic review of the efficacy of meditation techniques as treatments for medical illness. *Journal of Alternative & Complementary Medicine, 12*(8), 817-832.

Asante, K. W. (2000). Dance: Rhythmic forces, ancestral, p. An aesthetic analysis. Asmara, Eriteria: Africa World Press.

Asante, M. K. (1987). *The Afrocentric idea*. Temple University.

Asante, M. K. (2007). *An Afrocentric Manifesto, p. Toward an African renaissance. Polity.*

Asante, M. K. (2007). *An Afrocentric Manifesto: Toward an African renaissance*. Polity.

Asante, M. K. (2015). *The History of Africa: The Quest for Eternal Harmony* (2nd ed.). Routledge.

Asante, M. K., & Abarry, A. S. (1996). *African Intellectual Heritage: A Book of Sources*. Temple University.

Ascione, G. (2016). Decolonizing the 'Global': The Coloniality of Method and the Problem of the Unit of Analysis. *Cultural Sociology*, *10*(3), 317–334. doi:10.1177/1749975516644843

Au, K., & Jordan, C. (1981). Teaching reading to Hawaiian children: Finding a culturally appropriate solution cited in Ladson-Billings, G. (1995). Toward a theory of culturally relevant pedagogy. *American Educational Research Journal*, *32*(3), 465–491.

Awajiusuk, F. J. (2015). Prospects and Challenges of Teaching Religious Ethics. In C. B. Anyansi-Archibong (Ed.), *Contemporary Issues Surrounding Ethical Research Methods and Practice* (pp. 306–323). IGI Global. doi:10.4018/978-1-4666-8562-8.ch013

Bae C J., Douka K. & Petraglia M D. (2017). On the origin of modern humans: Asian perspectives. *Science*, *358*(6368). Doi:10.1126cience.aai9067

Bangura, A. K. (2012). From Diop to Asante, p. Conceptualizing and Contextualizing the Afrocentric Paradigm. *The Journal of Pan African Studies*, *5*(1), 103–125.

Barkin. (2003). Realist Constructivism. *International Studies Review*, *5*(3), 325-342.

Barnes, T. (2007). 'History has to play its role': Constructions of race and reconciliation in secondary school historiography in Zimbabwe, 1980-2002. *Journal of Southern African Studies*, *33*(7), 633–651. doi:10.1080/03057070701475740

Bassey, M. O. (2007). What is Africana critical theory or black existential philosophy? *Journal of Black Studies*, *37*(6), 914–935. doi:10.1177/0021934705285563

Batra, P. (2014). Problematising teacher education practice in India: Developing a research agenda. *Education as Change, 18*(sup1), S5-S18.

Batra, P. (2015). Curriculum in India. In Curriculum studies in India (pp. 35-63). Palgrave Macmillan.

Batra, P. (2019). Comparative Education in South Asia: Contribution, Contestation, and Possibilities. In Comparative and International Education: Survey of an Infinite Field (International Perspectives on Education and Society, Vol. 36). Emerald Publishing Limited. doi:10.1108/S1479-367920190000036011

Batra, P. (2005). Voice and agency of teachers: Missing link in national curriculum framework 2005. *Economic and Political Weekly*, 4347–4356.

Beale, M. A. (1998). *Apartheid and University Education, 1948 -1970.* Published Phd thesis: Obtained from. http://wiredspace.wits.ac.za/jspui/bitstream/10539/20610/1/Beale%20Mary%20Alice._Apartheid%20and%20university%20educati_001.pdf

Becken, H. J. (1968). On the Holy Mountain: A visit to the New Year's Festival of the Nazaretha Church on Mount Nhlangakazi, 14 January 1967. *Journal of Religion in Africa. Religion en Afrique, 1*(2), 138–149. doi:10.2307/1594789

Beckmann, J. (2016). Exclusion and Education in South Africa: An Education Law Perspective of Emerging Alternative Understandings of Exclusion. *Education Provision to Every One: Comparing Perspectives from Around the World BCES Conference Books, 14*(1), 295-302

Beckmann, J. (2016). Exclusion and Education in South Africa: An Education Law Perspective of Emerging Alternative Understandings of Exclusion. *Education Provision to Every One: Comparing Perspectives from Around the World BCES Conference Books, 14*(1), 295–302.

Behr, A. L., & Ngubentombi, S. V. S. (1988). *Education in South Africa.* Academica.

Ben-Jochannan, Y., & Clarke, J. H. (2017). *New dimensions in African history: the London lectures of Dr. Yosef ben-Jochannan and Dr. John Henrik Clarke / edited with introduction by John Henrik Clarke.* Brawtley.

Bernal, D. D. (2002). Critical race theory, Latino critical theory, and critical raced-gendered epistemologies: Recognizing students of color as holders and creators of knowledge. *Qualitative Inquiry, 8*(1), 105–126. doi:10.1177/107780040200800107

Bernasconi, R. (2001). With what must the philosophy of world history begin? On the racial basis of Hegel's eurocentrism. *Nineteenth-Century Contexts, 22*(2), 171–201. doi:10.1080/08905490008583507

Bernstein, B. (2000). *Pedagogy, symbolic control, and identity: Theory, research, critique* (Vol. 5). Rowman & Littlefield.

Berry, J. W. (1974). Psychological aspects of cultural pluralism: unity and identity reconsidered. *Topics in Culture Learning, 2*, 17-22.

Berry, L. L. (2010). What is the Hard Return on Employee Wellness Programs? *Harvard Business Review.* PMID:21188899

Bevan, S. (2006). *White farmer stands trial for shooting a black boy he 'mistook for a dog'.* Obtained at https://www.telegraph.co.uk/

Bhabha, H. (1994). The location of culture. New York, NY: Routledge.

Biko, S. (1998). *The African Philosophy Reader* (P. P. H. Coetzee & A. P. J. Roux, Eds.). Routledge.

Biko, S. (2004). *I write what I like: a selection of his writings/ Steve Biko* (C. R. Aelred Stubbs, D. Tutu, & N. Biko, Eds.). Picador Africa.

Binder, A. J. (Ed.). (2002). Contentious curricula. In Afrocentrism and creationism in American public schools. Princeton University Press.

Blaut, J. M. (1992). The theory of Cultural Racism. *A Radical. The Journal of Geography*, *23*, 289–299.

Boateng, F. (1983). African traditional education, p. A method of disseminating cultural values. *Journal of Black Studies*, 13.

Bonetto, S. (2006). Race and Racism in Hegel: An Analysis. *Minerva – An Internet. The Journal of Philosophy*, *10*, 35–64.

Bonilla-Silva. (2006). *Racism without racists: Color-blind racism and the persistence of racial inequality in the United States* (2nd ed.). Rowman & Littlefield Publishers, Inc.

Bouillion, L. M., & Gomez, L. M. (2001). Connecting school and community with science learning: Real world problems and school–community partnerships as contextual scaffolds. *Journal of Research in Science Teaching*, *38*(8), 878–898. doi:10.1002/tea.1037

Bourdieu, P., & Passeron, J.-C. (1990). Theory, culture & society. Reproduction in education, society and culture (2nd ed.; R. Nice, Trans.). Sage Publications, Inc.

Boyd, A., & Lenix-Hooker, C. J. (1992). Afrocentricism: Hype or history. *Library Journal*, *18*(117), 46–49.

Boyd, W. L. (1999). Paradoxes of Educational Policy and Productivity. *Educational Policy*, *13*(2), 227–250. doi:10.1177/0895904899013002001

Brain, J. B. (1990). *The Catholic Church in the Transvaal*. Missionary Oblates of Mary Immaculate.

Brazier, A., Mulkins, A., Sherman, A. C., & Thomas, P. G. (2001). *Research on Faith and Health: New Approach to Old Questions Psychological Perspectives*. Guilford Press.

Breidlid, A. (2013). Education, indigenous knowledges, and development in the global south; Contesting knowledges for a sustainable future. London: Routledge.

Brinkmann, S. (2015). Learner-centred education reforms in India: The missing piece of teachers' beliefs. *Policy Futures in Education*, *13*(3), 342–359. doi:10.1177/1478210315569038

Broad-based Black Economic Empowerment Act 53 of 2003.

Brown, H. D. (1987). *Principles of language learning and teaching*. Prentice-Hall, Inc.

Bruner, J. (1986). *Actual minds, possible worlds*. Harvard.

Bryant, K. D. (2018). Colonial education. In M. S. Shanguhyia & T. Falola (Eds.), The Palgrave handbook of African colonial and postcolonial history (pp. 281-302). New York: Palgrave MacMillan. doi:10.1057/978-1-137-59426-6_10

Burns, E. (2008). How Can a "Southern Theory" Perspective Contribute to New Zealand Counselling? *New Zealand Journal of Counselling*, *28*(2), 10–24.

Cabral, A. (1969). *Revolution in Guinea: An African People's Struggle*. Monthly Review Press.

Cajetas-Saranza, R. (2015). Approaches to Integrating Multicultural Perspectives in PNU-Mindanao: Basis for Multicultural Education Program. *International Peer Reviewed Journal, 16*, 37–55. doi:10.7718/iamure.ije.v16i1.1122

Caldwell, K., Emery, L., Harrison, M., & Greeson, J. (2011). Changing in Mindfulness, Wellbeing, and Sleep Quality in College Students Through Taijiquan Course: A Cohort Control Study. *Journal of Alternative and Complementary Medicine (New York, N.Y.), 17*(10), 931–938. doi:10.1089/acm.2010.0645 PMID:21999153

Cameroon, J., & Dodd, W. (1970). Society, Schools and Progress in Tanzania 1919-1970. London: Academic Press.

Canterbury Christ Church University (CCCU). (2014). *An introduction to ethics issues and principles in research involving human and animal participants*. CCCU.

Cassell, P. (1993). Problems of Action and Structure. A Giddens Reader. London: MacMillan. doi:10.1007/978-1-349-22890-4_3

Ce'saire, A. (2000). Discourse on Colonialism (J. Pinkham, Trans.). New York: Monthly Review Press.

Chan, C. L. W., Fan, F. W., & Gong, R. Y. (2003). *The Body-Mind-Spirit Integrative Health Approach: Group Counselling Theory and Application*. Ethnic Publishing. (in Chinese)

Chan, C. L. W., Ho, P. S. Y., & Chow, E. (2001). A body-mind-spirit model in health: An eastern approach. [PubMed]. *Social Work in Health Care, 34*(3-4), 261–282. doi:10.1300/J010v34n03_02

Chan, C., Ng, E., Chan, C., Ho, P. C., & Chan, T. (2006). Effectiveness of psychosocial group intervention for reducing anxiety in women undergoing in vitro fertilization: A randomized controlled study. *Fertility and Sterility, 85*(2), 339–346. doi:10.1016/j.fertnstert.2005.07.1310 PMID:16595209

Chavunduka, M. (1995). The missing links. keynote address to the workshop on the study and promotion of Indigenous knowledge systems and sustainable natural resources management in Southern Africa. *Midmar KwaZulu-Nata, l*, 24.

CHE (Council for Higher Education). (2017). Decolonising the curriculum: Stimulating debate. *Briefly Speaking, 3*, 1-12.

Chilisa, B. (2012). *Indigenous research methodologies*. Sage.

Chilisa, B., Major, T. E., & Khudu-Petersen, K. (2017). Research Methodologies, a postcolonial, African-based relational paradigm. *Qualitative Research, 17*(3), 326–339. doi:10.1177/1468794117696176

Chilisa, B., & Tsheko, G. N. (2014). Mixed Methods in Indigenous Research: Building Relationships for Sustainable Intervention Outcomes. *Journal of Mixed Methods Research, 8*(3), 222–233. doi:10.1177/1558689814527878

Chin, D. (1989). Interculturalism, postmodernism, pluralism. *Performing Arts Journal, 11*(3), 163–175. doi:10.2307/3245434

Chinese Association of South Africa v The Minister of Labour (2008) ZAGPHC 174.

Chou, L. (2005). *The Martial Artist's Book of Yoga: Improve Flexibility. Balance and Strength for Higher Kicks, Faster Strikes, Smoother Throws, Safer Falls, and Stronger Stances*. Academic Press.

Christie, P., & Collins, C. (1984). Apartheid Ideology and Labour Reproduction. In P. Kallaway (Ed.), *Apartheid and Education: The Education of Black South Africans*. Ravan Press.

Chubb, J., & Moe, T. (1990). *Politics, Markets and America's Schools*. Brookings Institution.

Chung, F. (2008). Accelerating the expansion of access to secondary education: The 1980 – 1990 experience in Zimbabwe. ADEA – 2008 Biennale on Education in Africa.

Clark, M.M., Warren, B.A., Hogen, M.D., Johnson, B.D., Jenkins, S.M., Wernerburg, M.S., & Olse, M.D. (2011). Stress Level, Health Behaviour, and Quality of Life in Employee Joining a Wellness Centre. *American Journal of Health Promotion, 26*(1), 21-25.

Classens, A. (2014). *Communal land, property rights and traditional leadership*. Paper presented at Wits Institute for Social and Economic Research.

Cloete, F., & Auriacombe, C. (2019). Revisiting decoloniality for more effective research and evaluation. *African Evaluation Journal, 7*(1). doi:10.4102/aej.v7i1.363

Cock, M., Dold, T., & Wiersum, F. (2003). The Cost of Culture. *Veld & Flora., 89*(4), 146–149.

Cockrell, A. (1996). Rainbow Jurisprudence. *South African Journal on Human Rights, 12*(1), 1–38. doi:10.1080/02587203.1996.11834900

Collins, A. K., & Millard, M. (2013). Transforming Education in South Africa: Comparative Perceptions of a South African Social Work Learning Experience. *Educational Review, 65*(1), 70–84. doi:10.1080/00131911.2011.648168

Connell, R. W. (1993). Schools and Social Justice. Temple University Press.

Connell, J. (2009). Dance Education: An examination of practitioners' perception in secondary schools and the necessity for teachers skilled in the pedagogy and content of dance. *Research in Dance Education, 10*(2), 115–130. doi:10.1080/14647890903019440

Connell, R. W. (1993). *Schools and Social Justice*. Temple.

Cox, T., & Nkomo, S. M. (1990). Invisible men and women, p. A status report on race as a variable in organisation behaviour research. *Journal of Organizational Behavior, 11*(6), 419–432. doi:10.1002/job.4030110604

Crush, J., & S Ramachandran, (2014). *Xenophobic violence in South Africa: Denialism,minimalism, realism*. Southern African Migration Project Policy Series 66. Cape Town: SAMP.

Darder, A. (2016). Foreword: Ruthlessness and the forging of liberatory epistemologies: An arduous journey. In J. M. Paraskeva (Ed.), Curriculum Epistemicide Towards an Itinerant Curriculum Theory. London: Routledge.

Dar, W. A., & Jan, K. (2017). Placing teacher at the junction of unity and diversity in India. *International Journal of Research in Engineering. IT and Social Sciences*, *7*(1), 56–59.

Dar, W. A., & Najar, I. A. (2017). Problems of Teaching Tribal Children—A Study on Tribal Schools of Khanshaib. *Academic Social Research*, *3*(4), 43–49.

Dar, W. A., & Najar, I. A. (2018). Educational Anthropology, Tribal Education and Responsible Citizenship in India. *South Asia Research*, *38*(3), 327–346. doi:10.1177/0262728018800759

Dastile, N.P., & Ndlovu-Gatsheni, S.J. (2013). Power, Knowledge and Being: Decolonial Combative Discourse as a Survival Kit for Pan-Africanists in the 21St Century, *Alternation*, *20*(1), 105 - 134.

De Houwer, J., & Moors, A. (2013). What is learning? On the nature and merits of a functional definition of learning. *Psychonomic Bulletin & Review*, *20*(4), 631–642. doi:10.375813423-013-0386-3 PMID:23359420

De Schutter, H. (2007). Nations beyond nationalism. *Inquiry*, *50*(4), 378–394. doi:10.1080/00201740701489468

De Wet, K. (2010). 2010, The Importance of Ethical Appraisal in Social Science Research: Reviewing a Faculty of Humanities' Research Ethics Committee. *Journal of Academic Ethics*, *8*(4), 301–314. doi:10.100710805-010-9118-8

De Wet, N. (2017). Family formation and economic insecurity among youth in South Africa. *Development Southern Africa*, *34*(6), 694–703. doi:10.1080/0376835X.2017.1310028

Dei, G. J. S. (2000). Rethinking the role of Indigenous knowledge in the academy. *International Journal of Inclusive Education*, *4*(2), 111–132. doi:10.1080/136031100284849

Dei, G. J. S. (2014). Indigenizing the curriculum: The case of the African university. In G. Emeagwali & G. S. Dei (Eds.), *African Indigenous Knowledge and the Disciplines* (pp. 165–180). Rotterdam Sense Publishers.

Delmas, A. (2011). From travelling to history: An outline of the VOC writing system during the 17th century. In A. Delmas & N. Penn (Eds.), *Written Culture in a Colonial Context: Africa and the Americas 1500–1900*. UCT Press. doi:10.26530/OAPEN_628137

Dempsey, P., & Dempsey, A. (1992). *Trials of War Criminals before the Nuremburg Military Tribunals: The Medical Case 1948*. Jones and Bartlett.

Department of Education. (2002). *Revised National Curriculum Statement Grades R-9 (Schools) Policy, Arts and Culture*. Government Printer.

Desai, K., & Sanya, B.N. (2016). Towards decolonial praxis: reconfiguring the human and the curriculum. *Gender and Education, 28*(6), 710–724. . doi:10.1080/09540253.2016.1221893

Dewey, J. (1956). *The school and society and the child and the curriculum.* University of Chicago Press.

Dixon, N. M. (1999). The changing face of knowledge. *The Learning Organization, 6*(5), 212–216. doi:10.1108/09696479910299929

Do Vale, H. F. (2016). Explaining Education Reforms and Decentralization in Brazil, South Africa, South Korea, and Spain. *Lex localis - Journal of Local Self-Government, 4*(3), 591–612. doi:10.4335/14.3.591-612(2016)

Dougall, J. W. C. (1963). *Christians in the African Revolution.* The Saint Andrew Press.

Douglas, H. P. (1979). The Passive Voice Theories in Religious Sociology. *Review of Religious Research, 21*(1), 51–61. doi:10.2307/3510155

Du Bois, W.E.B. (2016). *The Souls of the black folk.* Digireads.com Publishing. (Original publication 1903)

Durie, M. (2004). An Indigenous model of health promotion. *Health Promotion Journal of Australia, 15*(3), 181–185. doi:10.1071/HE04181

Duthely, M. L., Nunn, S. G., & Avella, J. T. (2019). Spirituality and Religion as Cultural Influences in Andragogy. In D. P. Peltz & A. C. Clemons (Eds.), *Multicultural Andragogy for Transformative Learning* (pp. 45–72). IGI Global. doi:10.4018/978-1-5225-3474-7.ch004

Elder, C., & Potskin, J. (2018). Interdisciplines, and Indigenous research and methodologies. In C. Lury, R. Fensham, A. Heller-Nicholas, S. Lammes, A. Last, M. Michael, & E. Uprichard (Eds.), Routledge handbook of interdisciplinary research methods (pp. 311-315). New York: Routledge. doi:10.4324/9781315714523-43

Ellsworth, E. (1997). Teaching positions. Difference, pedagogy, and the power of address. New York: Teachers College Press.

Emeagwali, G. (2003). African indigenous knowledge systems (AIK): Implications for the curriculum. In T. Falola (Ed.), *Ghana in Africa and the world: Essays in honour of AduBoahen.* Africa World Press.

Fafunwa, B. (1974). *A History of Education in Nigeria.* Allen and Unwin.

Falola, T. (2005). Writing and teaching national history in an era of global history. *Africa Spectrum, 40*(3), 499–519.

Fanelli, D. (2009). *How Many Scientists Fabricate and Falsify Research? A Systematic Review and Meta-Analysis of Survey Data.* doi:10.1371/journal.pone.0005738

Fanon, F. (1965). A dying colonialism (H. Chevalier, Trans.). New York: Monthly Review Press.

Fanon, F. (1967). The wretched of the earth. New York: Grove Press.

Fanon, F. (2001). The wretched of the earth (C. Farrington, Trans.). London: Penguin. (Original publication 1961)

Fanon, F. (2008). Black skin, white masks (C. L. Markmann, Trans.). London: Pluto. (Original publication 1952)

Fanon, F. (2008). Black skin, white masks (C.L. Markmann, Trans.). London: Pluto. (Original publication 1952)

Fanon, F. (1964). *Towards the African Revolution*. Grove Press.

Fanon, F. (1968). *The Wretched of the Earth*. Grove Press.

Fernandez, J. W. (1973). The precincts of the Prophet: A day with Johannes Galilee Shembe. *Journal of Religion in Africa. Religion en Afrique*, *5*(1), 32–53. doi:10.1163/157006673X00041

Finkelstein, C., Brownstein, A., Lan, Y. L., & Scott, C. (2007). Anxiety and Stress Reduction in Medical Education: An Intervention. *Medical Education*, *41*(3), 258–264. doi:10.1111/j.1365-2929.2007.02685.x PMID:17316210

Flavier, J. M. (1995). The regional program for the promotion of indigenous knowledge in Asia. In D. M. Warren, L. J. Slikkerveer, & D. Brokensha (Eds.), *The cultural dimension of development: Indigenous knowledge systems* (pp. 479–487). Intermediate Technology Publications. doi:10.3362/9781780444734.043

Formicola, V., & Holt, B. M. (2015). Tall guys and fat ladies: Grimaldi's Upper Paleolithic burials and figurines in an historical perspective. *Journal of Anthropological Sciences*, *93*, 71–88. doi:10.4436/jass.93007 PMID:25992636

Foucault, M. (1980). Power/Knowledge: Selected interviews and other writings 1972–1977. New York: Pantheon.

Freire, P. (1972). *Pedagogue of the oppressed*. Penguin.

Freire, P. (1979). *Pedagogy of the Heart*. Continuum.

Freire, P. (1990). *Pedagogy of the oppressed.* New York: *Continuum.*

Fúnez-Flores, J. I., & Phillion, J. (2019). A Political ontological approach to decolonization of ethnographic research in education. In Rethinking 21st century diversity in teacher preparation, K-12 education, and school policy (pp. 39-54). Cham: Springer. doi:10.1007/978-3-030-02251-8_3

Gade, C. (2012). What is ubuntu? Different interpretations among South Africans of African descent. *South African Journal of Philosophy*, *31*(3), 484–504. doi:10.1080/02580136.2012.10751789

Garriott, P. O., Reiter, S., & Brownfield, J. (2016). Testing the Efficacy of Brief Multicultural Education Interventions in White College Students. *Journal of Diversity in Higher Education*, *9*(2), 158–169. doi:10.1037/a0039547

Gatsheni, N. (2013). *Coloniality of Power in Postcolonial Africa Myths of Decolonization*. Academic Press.

Gbolonyo, J. S. K. (2009). *Indigenous Knowledge And Cultural Values In Ewe Musical Practice: Their Traditional Roles And Place In Modern Society* (PhD dissertation). University Of Pittsburgh.

Gelling, L., & Munn-Giddings, C. (2011). Ethical review of action research: The challenges for researchers and research ethics committees. *Research Ethics Review*, *7*(3), 100–106. doi:10.1177/174701611100700305

Gemma, B., Lesikar, O. N., & Yunus, R. (2003). *Education, indigenous knowledge and globalisation*. Aang Serian Community College.

Ghandi, L. (1998). Postcolonial theory. In A Critical Introduction. Edinburgh University Press.

Giambrone, C.A., Cook-Cottone, C.P., & Klein, J.E. (2018). *The Africa Yoga Project and Well-Being: A Concept Map of Students' Perceptions*. doi:10.1111/aphw.12124

Gibson, J., & McDonald, H. (2001). *Truth-yes, reconciliation-maybe: South Africans judge the truth and reconciliation process*. Institute for Justice and Reconciliation.

Giliomee, H., & Mbenga, B. (2007). *New history of South Africa*. Tafelberg.

Gittens, A.F. (2012). Black dance and the fight for flight: Sabar and the transformation and cultural significance of dance from West Africa to Black America (1960-2010). *Journal of Black Studies*, *43*(1), 49-71.

Gonye, J. (2015). *Representations of dance in Zimbabwean literature, post 1960* (Unpublished doctoral thesis). University of Venda.

Gonye, J., & Moyo, N. (2018). African Dance as an Epistemic Insurrection in Postcolonial Zimbabwean Arts Education Curriculum. In The Palgrave handbook of race and the arts in education (pp 157-174). Cham: Palgrave Macmillan. doi:10.1007/978-3-319-65256-6_9

Gonye, J. (2013). Mobilizing dance/traumatizing dance: Kongonya and the politics of Zimbabwe. *Dance Research Journal*, *45*(1), 65–79. doi:10.1017/S0149767712000277

Gonye, J., & Moyo, N. (2015). Traditional African dance education as curriculum reimagination in post-colonial Zimbabwe: A re-think of policy and practice of dance education in the primary schools. *Research in Dance Education*, *16*(3), 259–275. doi:10.1080/14647893.2015.1036020

Gonzalez, N., & Moll, L. C. (2002). Cruzando el puente: Building bridges to funds of knowledge. *Educational Policy*, *16*(4), 623–641. doi:10.1177/0895904802016004009

González, N., Neff, D., Amanti, C., & Moll, L. (2006). Funds of knowledge for teaching: Using a qualitative approach to connect homes and classrooms. In *Funds of knowledge* (pp. 83–100). Routledge. doi:10.4324/9781410613462

Gordon, B. M. (1990). The Necessity of African-American Epistemology for Educational Theory and Practice1. *Journal of Education, 172*(3), 88–106. doi:10.1177/002205749017200307

Gordon, L. R. (1995). *Bad faith and anti-black racism.* Humanities Press.

Gordon, L. R. (2006). *Disciplinary decadence. Living thought in trying times.* Paradigm.

Gottschild, B. D. (2003). The black dancing body, a geography from Coon to Cool. New York: Palgrave. doi:10.1007/978-1-137-03900-2

Graybill, L. (2002). *Truth and reconciliation in South Africa.* Lynne Rienner.

Greene, M. (1995). Releasing the imagination, essays on education, the arts, and social change. San Francisco, CA: Jossey-Bass.

Gregory, G. J., Luke, A., & Green, B. (2008) Introduction: What counts as knowledge in educational settings, p. Disciplinary knowledge, assessment, and curriculum. Review of Research in Education, 32, p. vii–x. Doi:10.3102/0091732X07311063

Griffin, S. F. (1995). 'Shembe' is the way. Durban: Institute of Current World Affairs (ICWA).

Grosfoguel, R. (2007). The Epistemic Decolonial Turn. *Cultural Studies, 21*(2-3), 211–223. doi:10.1080/09502380601162514

Gruenewald, D. A. (2003). The best of both worlds: A critical pedagogy of place. *Educational Researcher, 32*(4), 3–12. doi:10.3102/0013189X032004003

Guilhermina, M. L. (2007). The Limits and Possibilities of ICT in Education. *Sísifo: Educational Sciences Journal, 3*, 39–48.

Hardt, M., & Negri, A. (2000). Empire. Harvard University Press.

Hartshorne, K. (1992). *Crisis and challenge: Black education 1910-1990.* Oxford University Press.

Haydock, K. (2015). Stated and unstated aims of NCERT social science textbooks. *Economic and Political Weekly, 50*(17), 109–119.

Heeren, A. H. L. (1832). *Reflections on the politics, intercourse and trade of the ancient nations of Africa.* Oxford: D.A. Taalboys.

Hegde, R. (2018). Curriculum under Siege: Yesterday Once More? *Contemporary Education Dialogue, 15*(2), 217–226. doi:10.1177/0973184918781218

Heilman, D. C., & Clarke, D. L. (2016). Transformative Learning Theory: Perspectives on Nelson Mandela and Application for U.S. Learners in South Africa. In S. M. Tomlinson-Clarke & D. L. Clarke (Eds.), *Social Justice and Transformative Learning: Culture and Identity in the United States and South Africa* (pp. 40–57). Routledge.

Heleta, S. (2016). Coloniality persist in our universities and we must urgently decolonise. *Mail & Gaidian*. Obtained at https://mg.co.za/article/2016-11-18-00-coloniality-persists-in-our-universities-and-we-must-urgently-decolonise

Herrera, L. L. (2015). *Meridian Systems Yoga: An Innovative & Accessible Method That Combines Yoga & Traditional Chinese*. Medicine Create Space Independent Publishing Platform.

Higher Education South Africa. (2014). *South African Higher Education in the 20th Year of Democracy: Context, Achievements and Key Challenges HESA presentation to the Portfolio Committee on Higher Education and Training Cape Town*. Retrieved October 17, 2016, from https://pmg.org.za/committeemeeting/17037/

Higher Education South African. (2014). *South African Higher Education in the 20th Year of Democracy: Context, Achievements and Key Challenges HESA presentation to the Portfolio Committee on Higher Education and Training Cape Town*. Retrieved October 17, 2016, from https://pmg.org.za/committeemeeting/ 17037/

Howe, M., Leidel, A., Krishnan, S. M., Weber, A., Rubenfire, M., & Jackson, A. E. (2010). Patient Related Diet and Excise Counselling: Do Providers' Own Lifestyle Habits Matters? *Preventive Cardiology*, *13*(4), 180–185. doi:10.1111/j.1751-7141.2010.00079.x PMID:20860642

Hymes, D. (1974). *Foundations in Sociolinguistic: An Ethnographic Approach*. University of Pennsylvania.

Hyslop, J. (1989). A Destruction Coming In Bantu Education as a Response to Social Crisis. African Studies Institute.

Idang, G. E. 2015. African culture and values. *African Culture and Values*, *16*(2), 97-111.

Idang, G. E. (2009). The Mind-Body problem in African Culture. In *From Footmarks to Landmarks on African Philosophy*. Obaroh and Ogbinaka.

Ignas, V. (2004). Opening doors to the future: Applying local knowledge in curriculum development. *Canadian Journal of Native Education*, *28*(1/2), 49.

Ikeda, A. (2011). Optimism in Relation to Inflammation and Endtheial Dysfunction on Older Men: The VA Normative Aging Study. *Psychosomatic Medicine*, *73*, 664–671. doi:10.1097/PSY.0b013e3182312497 PMID:21949417

Ivtzan, I., Lomas, T., & Hefferon, K. (2014). *Applied Positive Psychology: Integrated Positive Practice*. Sage Publications.

Ivtzan, I., & Papantoniou, A. (2014). Yoga meets positive psychology: Examining the integration of hedonic (gratitude) and eudaimonic (meaning) wellbeing in relation to the extent of yoga practice. *Journal of Bodywork and Movement Therapies*, *18*(2), 183–189. doi:10.1016/j.jbmt.2013.11.005 PMID:24725784

Jadhav, S. G. U., & Havalappanavar, N. B. (2009). Effects of yoga interventions on anxiety and subjective well-being. *Journal of the Indian Academy of Applied Psychology, 35*, 27-31.

James, M. (1993). *Health Essentials Chi Kung Cultivate Personal Energy*. Element.

Jenks, C. (1993). *Culture*. Routledge.

Johns, S. (2004). Isaiah Shembe and the Uses of Textuality. *Journal of Southern African Studies, 30*(2), 401–403.

Kallaway, P. (2002). The History of Education under Apartheid, 1948–1994: The Doors of Learning and Culture Shall be Opened. In P. Kallaway (Ed.), *History of Schools and Schooling* (pp. 224–242). Peter Lang.

Kallaway, P. (2002). *The history of education under apartheid, 1948–1994: the doors of learning and culture shall be opened*. Pearson.

Kallaway, P. (Ed.). (1990). *Apartheid and education: the education of black South Africans* (4th ed.). Ravan Press.

Kaminoff, L., & Mathews, A. (2012). *Yoga Anatomy. Your illustrated guide to postures, movements, and breathing techniques*. Human Kinetics.

Karabel, J. (1996). Towards a Theory of Intellectuals and Politics. *Theory and Society, 25*(2), 205–233. doi:10.1007/BF00161141

Kaya, H. O., Kamwendo, G. H., & Rushubirwa, L. (2016). African Indigenous Languages in Higher Education. *Studies of Tribes and Tribals, 14*(2), 121–128. doi:10.1080/0972639X.2016.11886739

Kaya, H., & Seleti, Y. (2013). African indigenous knowledge systems and relevance of higher education in South Africa. *International Education Journal: Comparative Perspectives, 12*(1), 30–44.

Kent, H. (2000). *Yoga An Introduction Guide to Optimum Health for Mind, Body and Spirit*. Academic Press.

Khalsa, S. B., Hickey-Schultz, L., Cohen, D., Steiner, N., & Cope, S. (2012). Evaluation of the Mental Health Benefit of Yoga in Secondary School: A Preliminary Randomized Controlled Trail. *The Journal of Behavioral Health Services & Research, 39*(1), 80–90. doi:10.100711414-011-9249-8 PMID:21647811

Khorommbi, N. L. (1996). *Echoes from beyond a pass between two mountains* (Unpublished Master of Theology dissertation). University of South Africa.

Khoza-Shangase, K. (2019). Intellectual and Emotional Toxicity: Where a Cure Does Not Appear to Be Imminent. In G. Khunou, H. Canham, K. Khoza-Shangase, & E. D. Phaswana (Eds.), Black Academic Voices: The South Africa Experiences. Academic Press.

Khunou, G. (2019). Black in the academy: Reframing knowledge, the knower, and know. In G. Khunou, H. Canham, K. Khoza-Shangase, & E. D. Phaswana (Eds.), Black Academics Voices: The South African Experience. Academic Press.

Kiguwa, P. (2019). Negotiating the academy: Black bodies 'out of place. In G. Khunou, H. Canham, K. Khoza-Shangase, & E. D. Phaswana (Eds.), Black Academic Voices: The South Africa Experiences. Academic Press.

Killingsworth, M. A., & Gilbert, D. T. (2010). A Wondering Mind is un Happy Mind. *Science, 330*(6006), 932. doi:10.1126cience.1192439 PMID:21071660

Kim, S. Y. H., Ubel, P., & De Vries, R. (2009). *Pruning the regulatory tree.* PubMed., doi:10.1038/457534a

Kincheloe, J. L. (2007). Postformalism and critical ontology- part 1: Difference, indigenous knowledge, and cognition. In J. L. Kincheloe & R. A. Horn (Eds.), The Praeger handbook of education and psychology Vol 4. Westport: Greenwood Publishing Company.

Kincheloe, J. L. (2011). Critical ontology: Visions of selfhood and curriculum. In K. Hayes, S. Steinberg, & K. Tobin (Eds.), Key work in critical pedagogy (pp. 201–217). Rotterdam: Sense Publishers.

Klein, C. (2003). Cold war orientalism; Asia in the Middlebrow imagination, 1945-1961. Berkeley: University of California Press.

Knapik, J. J., Steelman, R. A., Hoedebecke, S. S., Austin, K. G., Farina, E. K., & Lieberman, H. R. (2016). Prevalence of dietary supplement use by athletes: Systematic review and meta-analysis. *Sports Medicine (Auckland, N.Z.), 46*(1), 103–123. doi:10.100740279-015-0387-7 PMID:26442916

Knappert, J. (1995). *An encyclopedia of myth and legend: African mythology.* Diamond.

Kori, E. (2016). Challenges to academic freedom and institutional autonomy in South African universities. *International Journal of Teaching and Education, IV*(1), 45–53. doi:10.20472/TE.2016.4.1.004

Kritzinger, J. J. (1988). *The South African context/or mission.* Lux Verbi.

Kruger, E. G. (Ed.). (1986). *Education: past, present and future.* Euro Publications.

Kuhn, T. (1970). *The structure of scientific revolutions* (2nd ed.). University of Chicago Press.

Kumar, K. (1983). Educational experience of scheduled castes and tribes. *Economic and Political Weekly*, 1566–1572.

Kumar, K. (1986). Textbooks and educational culture. *Economic and Political Weekly*, 1309–1311.

Kumar, K. (1988). Origins of India's" textbook Culture. *Comparative Education Review, 32*(4), 452–464. doi:10.1086/446796

Kumar, R. (2005). *Research Methodology: A Step-by-Step Guide for Beginners.* Sage Publications.

Kumar, V. (2016). How Egalitarian Is Indian Sociology? *Economic and Political Weekly, 51*(25), 33–39.

Kunhiyop, S. W. (2008). *African Christian Ethics.* WordAlive.

Kuwor, S. K. (2003). *Transmission of Anlo-Ewe Dances in Ghana and in Britain: Investigating, reconstructing and disseminating knowledge embodied in the music and dance traditions of Anlo-Ewe people in Ghan.* (Unpublished doctoral thesis). Department of Dance Studies, University of Roehampton.

Ladson-Billings, G. (1995). Toward a theory of culturally relevant pedagogy. *American Educational Research Journal, 32*(3), 465–491. doi:10.3102/00028312032003465

Ladson-Billings, G. (2000). Racialized discourses and ethnic epistemologies. In N. K. Denzin & Y. S. Lincoln (Eds.), *Handbook of qualitative research* (2nd ed., pp. 257–277). Sage.

Ladzani, K. Y. (2014). *Moral regeneration in the lives of Vhavenda youth through indigenous knowledge systems: applied ethnography of communication-based approaches with special reference to Tshivenda.* Unpublished thesis.

Lall, M. (2008). Educate to hate: The use of education in the creation of antagonistic national identities in India and Pakistan. *Compare: A Journal of Comparative Education, 38*(1), 103–119. doi:10.1080/03057920701467834

Lalvani, V. (1999). *Yoga for Stress: Relax and re-energize the natural yoga way.* Octopus Publishing Group.

Lalvani, V. (2000). *Complete book of Yoga. The total yoga workout for mind body and spirit.* Octopus Publishing Group.

Landman, W., Killian, C., Swanepoel, E., & Bodenstein, H. (1982). *Fundamental Pedagogics.* Juta.

Lave, J. (1988). *The culture of acquisition and the practice of understanding.* IRL Report 88-0008, Institute for Research on Learning.

Le Roux, E.H. (2013). *Between Complicity and Resistance: A Social History of the University Presses in Apartheid South Africa.* Published Phd thesis: Obtained from https://repository.up.ac.za/bitstream/handle/2263/25429/00front.pdf

Lengrand, P. (1970). *An introduction to lifelong education.* UNESCO.

Leonardo, Z. (2018). Dis-orienting western knowledge coloniality, curriculum and crisis. *The Cambridge Journal of Anthropology, 36*(2), 7–20. Doi:10.3167/cja.2018.360203

Lephalala, M. M. K. (2012). *Between Past and Present: Conflict in schools, an Ubuntu Perspective.* University of South Africa Press.

Letseka, M. (2012). In Defense of *Ubuntu*, Studies in Philosophy and Education. *International Journal (Toronto, Ont.), 31*(1), 47–60.

Letsekha, T. (2013). Revisiting the debate on the Africanisation of higher education: An appeal for a conceptual shift. *The Independent Journal of Teaching and Learning, 8*, 5–18. https://www.edge.org/conversation/lera_boroditsky-how-does-our-language-shape-the-way-we-think

Lewis, A. (1992). *Individual and community needs as aspects of the education crisis. A historical educational overview* (Unpublished MEd dissertation). University of Stellenbosch.

Lincoln, Y.S., & Guba, E.G. (1995). Establishing trustworthiness. *Naturalistic Inquiry, 289*, 331.

Lorde, A. (1984). The Master's Tools Will Never Dismantle the Master's House. In Sister Outsider. Crossing Press.

Louis, R. P. (2006). Can you hear us now? Voices from the margin: Using indigenous methodologies in geographic research. *Geographical Research*, *45*(2), 130–139. doi:10.1111/j.1745-5871.2007.00443.x

Lourens, A. C. U., Viljoen, A. M., & van Heerden, F. R. (2008). South African Helichrysum species: A review of the traditional uses, biological activity and phytochemistry. *Journal of Ethnopharmacology*, *119*(3), 630–652. doi:10.1016/j.jep.2008.06.011 PMID:18606217

Mabingo, A. (2019). Dancing with the "other": Experiential and reflective learning of African neo-traditional dances through dance education study abroad. International Journal of Education & the Arts, 20(1). doi:10.18113/P8ijea20n2

Mabunda, D. C. (1995). *A historico-educational survey and evaluation of Swiss Missionary Education at Lemana* (Unpublished MEd dissertation). Unisa.

Makgoba, W. (Ed.). (1999). African renaissance. Mafube Publishing.

Malis, M., & Missedja, T.Q. (2019). Schooled for servitude: The education of African children in British colonies, 1910–1990. *Genealogy, 3*(40), 1-12. doi:10.3390/genealogy3030040

Mangena, F. (2012). Towards a hunhu/ubuntu dialogical moral theory. *Phronimon, 13*(2), 2012.

Mashau, T. D. (2004). The Reformed Churches in South Africa's missionary pilgrimage among the Vhavenda people at Siloam. *Studia Historiae Ecclesiasticae*, *30*(1), 90–107.

Mashau, T. D. (2007). Occultism in an African context, p. a case for the Vhavenda-speaking people of the Limpopo Province. *In die Skriflig*, *41*(4), 637–653. doi:10.4102/ids.v41i4.324

Mason, J. (1994). Linking Qualitative and Quantitative data analysis. In A. Bryman & R. G. Burgess (Eds.), *Analyzing Qualitative Data*. Routledge. doi:10.4324/9780203413081_chapter_5

Mawasha, A. L. (1969). *From Missionary and provincial control to state and community control of Bantu education: an appraisal* (Unpublished MEd dissertation). Unisa.

Mbeki, T. (2005). *Goals of Higher Education in Africa*. Retrieved from www.utexas.edu/conferences/africa

Mbeki, T. (1998). *Africa: The time has come.* Tafelberg.

Mbembe, A. J. (2016). Decolonizing the university: New directions. *Arts and Humanities in Higher Education*, *15*(1), 29–45. doi:10.1177/1474022215618513

Mbigi, L. (1995). *Ubuntu: A Rainbow Celebration of Cultural Diversity*. Ubuntu School of Philosophy,Tshwane.

Mbigi, L., & Maree, J. (1995). *Ubuntu: The Spirit of African Transformation Management*. Sigma Press.

McCulloch, J. (1983). *Black Soul, White Artifact. Fanon's clinical psychology and social theory*. Cambridge University Press. doi:10.1017/CBO9780511558559

McIntyre, E., Rosebery, A. S., & González, N. (Eds.). (2001). *Classroom diversity: Connecting curriculum to students' lives* (Vol. 88). Heinemann.

McLaren, P. L., & Giroux, H. A. (1990). Critical pedagogy and rural education: A challenge from Poland. *Peabody Journal of Education*, *67*(4), 154–165. doi:10.1080/01619569009538705

McLaughlin, B. W., Wang, D., Noone, A. M., Liu, N., Harazduk, N., Lumpikin, M., Harmati, A., & Saunders, P. (2011). Stress Biomarkers in Medical Student Participating in a Mind Body Medicine Skills Program. *Evidence-Based Complementary and Alternative Medicine*, *2011*, 1–8. doi:10.1093/ecam/neq039

McPhail, G. (2016). The fault lines of recontextualisation: The limits of constructivism in education. *British Educational Research Journal*, *42*(2), 294–313. doi:10.1002/berj.3199

Medina, J. (2011). Toward a Foucaultian epistemology of resistance: Counter-memory, epistemic friction, and guerrilla pluralism. *Foucault Studies*, *12*, 9–35. doi:10.22439/fs.v0i12.3335

Meersman, B. (2012). Soaring search for truth. *Mail and Guardian*. Retrieved: September, 2012, from www.mg.co.za/print/2012-05-04-soaring-search-for-truth

Melan, C. G. (2004). *Storytelling in Teaching*. The Association for Psychological Science.

Merriam, E. O. (2007). *Learning in Adulthood: A Comprehensive Guide*. Jossey- Bas.

Meyer, A. (2012). Suid-Afrika se oerverlede. Geskiedenis van Suid-Afrika.

Mezirow, J. (2009). Transformative Learning Theory. In J. Mezirow & E. W. Taylor (Eds.), *Transformative Learning in Practice: Insights from Community, Workplace, and Higher Education* (pp. 18–32). Jossey Bass.

Mignolo, W. (2009). Epistemic disobedience, independent thought and decolonial freedom. *Theory, Culture & Society*, *26*(7–8), 159–181. doi:10.1177/0263276409349275

Miller, W. R. (1999). *Integrating Spirituality into Treatment: Resources for Practitioners*. American Psychological Association. doi:10.1037/10327-000

Ministry of Primary and Secondary Education: Curriculum Framework for Primary and Secondary Education 2015-2022. (n.d.). Harare: Curriculum Development and Technical Services.

Ministry of Primary and Secondary Education: Dance Syllabus Forms 1 – 4. (n.d.). Harare: Curriculum Development and Technical Services.

Ministry of Primary and Secondary Education: History Syllabus Forms 1 – 4. (n.d.). Harare: Curriculum Development and Technical Services.

Minnaar, A., Offringa, D., & Payze, C. (1992). *To live in fear: Witch burning and medicine murder in Venda*. HSRC.

Mishra, R. C. (2019). Cultural and Psychological Dimensions of Development of the Adivasi Communities. In J. Pandey, R. Kumar, & K. Thapa (Eds.), *Psychological Perspectives on Diversity and Social Development*. Springer. doi:10.1007/978-981-13-3341-5_7

MisirHiralall, S. D. (2014). The postcolonial reality of using the term "liturgical" to describe Hindu dance. *Journal of Research on Christian Education*, *23*(2), 154-175. DOI:. doi:10.1080/10656219.2014.899479

Mkabela, Q. (2005). Using the Afrocentric method in researching indigenous African culture. *Qualitative Report*, *10*(1), 178–189.

Mlatsheni, C. (2012). The challenges unemployment imposes on youth. In H. Perold, N. Cloete, & J.Papier (Eds.), Shaping the future of South Africa's youth: Rethinkingpost-school education and skills training (pp. 31-41). Cape Town: Centre for Higher Education Transformation (CHET).

Mndende, N. (2009). Tears of the Distress: Voices of a Denied Spirituality in a Democratic South Africa. Dutywa: Icamagu Institute.

Mohanty, C. (1984). Under western eyes: Feminist scholarship and colonial discourses. *Boundary*, *12*(3).

Moll, L., Amanti, C., Neff, D., & Gonzalez, N. (2005). Funds of knowledge for teaching: Using a qualitative approach to connect homes and classrooms. *Funds of knowledge: Theorizing practices in households, communities, and classrooms*, 71-87.

Moumouni, A. (1968). *Education in Africa*. Praeger.

Moyana, T. T. (1989). Education, liberation, and the creative act. Harare:. Zimbabwe Publishing House.

Moyo, N. (2014). Nationalist history, nation-state making and secondary school history, p. Curriculum policy in Zimbabwe 1980-2010. *NORDIDACTICA Journal of Humanities and Social Science Education,* 1-24. Available at www.kau.se/nordidactica

Moyo, N., & Modiba, M. (2013). Who does this history curriculum want you to be? p. Representation, school history and curriculum in Zimbabwe. *Yesterday and Today*, *10*, 1–24.

Mpanza, M. P. (1999). *Submission to the Truth and Reconciliation Commission by Ibandla Lama Nazareth*. Retrieved: June, 2011, from web.uct.ac.za/depts/ricsa/commiss/trc/nazar.htm

Mpofu, E. (2002). Indigenization of the psychology of human intelligence in SubSaharan Africa. In W.J. Lonner, D.L. Dinnel, S.A. Hayes, & D.N. Sattler (Eds.*), On-line Readings in Psychology and Culture*. Center for Cross-Cultural Research, Western Washington University.

Mpofu, E. (1994). Exploring the self-concept in an African culture. *The Journal of Genetic Psychology, 155*(3), 341–354. doi:10.1080/00221325.1994.9914784 PMID:7964660

Msila, V. (2015). Ubuntu, Shaping the current workplace with (African) wisdom. Knowres Publishing (Pty) Ltd.

Mthembu, N. (2009). *The State of Current African Borders and their Rationale: A Case of Recent Violence in Azania [South Africa] against Black African Populace*. Paper presented at the annual SASA Congress, University of the Witwatersrand, Johannesburg.

Mthembu, N. (2016). Youth Challenges in the Transition to Adulthood in a South African Township. *Youth Voice Journal Online, 6*. Retrieved December 18, 2017, from https://youthvoicejournal.com/2016/12/

Mthembu, N. (2016). Youth challenges in the transition to adulthood in a South African township. Youth Voice Journal.

Mthembu, N. C. (2017). *Black African Township youth survival strategies in post-apartheid South Africa: A case study of the KwaMashu Township within eThekwini Municipality* (PhD Thesis). University of South Africa, Tshwane.

Mthembu, N. C. (2017). *Black African Township Youth Survival Strategies in Post-apartheid South Africa: A Case Study of the KwaMashu Township within eThekwini Municipality* (PhD Thesis). University of South Africa, Tshwane.

Mthembu, N. (2011). *Indigenous Knowledge creation Methodology*. Lambert.

Mthembu, N. (2018). Ushering Transformative Change: Infusing De-colonized Cultural Changes into South African Curriculum and Andragogical Practices. In D. P. Peltz & A. C. Clemons (Eds.), *Multicultural Andragogy for Transformative Learning* (pp. 194–206). IGI Global.

Mudimbe, V. Y. (1988). *The invention of Africa: Gnosis, Philosophy, and the order of knowledge*. Indiana University Press.

Muller, C. A., & Buchanan, D. A. (1994). *Nazarites song, dance, and dreams: the sacralization of time, space and the female body in South Africa.* New York: UMI Dissertation Services.

Mushi P. A K. (2009). *History of Education in Tanzania*. Dar-es-Salaam: Dar-esSalaam University.

Nabudere, D. W. (2011). *Afrikology, Philosophy and Wholeness: An Epistemology*. Africa Institute of South Africa.

National Science Foundation. (n.d.). *What is Interdisciplinary Research?* nsf.gov/od/oia/additional_resources/interdisciplinary_research/definition's

Ndebele, P., Mwaluko, G., Kruger, M., Ouwe, O., Oukem-Boyer, M., & Zimba, M. (2014). History of Research Ethics Review in Africa. In Research Ethics in Africa: A Resource for Research Ethics committees. Sun MeDIA Stellenbosch.

Ndhala, A. R., & Staden, J. V. (2012). Smokescreens and mirrors in safety and quality of herbal medicines: A case of commercialized herbal preparations. *South African Journal of Botany, 82*, 4–10. doi:10.1016/j.sajb.2012.06.008

Ndlovu-Gatsheni S.J., (n.d.). Global Coloniality and the challenges of crating African futures. *Strategic Review for Southern Africa, 36*(2).

Ndlovu-Gatsheni, S. J. (2013). Coloniality of power in postcolonial Africa: Myths of decolonization. Dakar: CODESRIA Book Series.

Ndlovu-Gatsheni, S. J. (2013a). Empire, global coloniality, and Arican objection. Berghahn Books.

Ndlovu-Gatsheni, S.J. (2017). *The emergence and trajectories of struggles for an 'African university': The case of unfinished business of African epistemic decolonization*. Academic Press.

Ndlovu-Gatsheni, S. J. (2013). *Coloniality of Power in Postcolonial Africa Myths of Decolonisation*. CODESRIA.

Ndlovu-Gatsheni, S. J. (2015). Genealogies of coloniality and implications for Africa's development. *Africa Development. Afrique et Developpement, 40*(3), 13–40.

Ndlovu-Gatsheni, S. J. (2018). Genealogies of Coloniality and Implications for Africa's Development. In V. Gumede (Ed.), *Inclusive development in Africa: Transformation of Global Relations*. doi:10.2307/j.ctvh8r26b.9

Ndlovu, S. M. (2006). *The Road to Democracy in South Africa*. Academic Press.

Newman, J. (1982). *Foundations of religious tolerance*. University of Toronto. doi:10.3138/9781487577759

Ngara, C. (2007). African ways of knowing and pedagogy revisited. *Journal of Contemporary Issues in Education, 2*(2), 7–20.

Nkomo, M. S. (2011). A postcolonial and anti-colonial reading of 'African' leadership and management in organization studies: tensions, contradictions and possibilities. *Organization, 18*(3), 365–386. doi:10.1177/1350508411398731

Nkomo, S. (1992). The emperor has no clothes, p. Rewriting 'race into the study of organizations'. *Academy of Management Review, 17*(3), 487–513.

Nkomo, S. M. (2006). *Images of African Leadership and Management in Organisation Studies: Tensions, Contradictions and Re-visions*. University of South Africa.

Nkondo, M. (2012). *Indigenous African knowledge systems in a polyepistemic world: The capabilities approach and the translatability of knowledge systems*. Paper presented at the Southern African Regional Colloquium on Indigenous African knowledge systems methodologies and epistemologies for research, teaching, learning and community engagement in higher education. Howard College Campus: University Of KwaZulu-Natal.

Nkrumah, K. (1965). *Neo-Colonialism, the Last Stage of Imperialism*. Thomas Nelson and Sons.

Nnabugwu, M.B., & Nwanegbo, J. (2012). Dysfunctional state and severance in Africa, p. an analysis in specificities and contradictions. *Ogirisi: A New Journal of African Studies, 9*, 127-140.

Nomadlala, N, Assounga, A., & Hariparshad, S. (2012). *Kidney and Diseases in an African Setting*. Reach Publishers.

Nsamenang, A. B. (2005). Educational development and knowledge flow: Local and global forces in human development in Africa. *Higher Education Policy, 18*(3), 275–288. doi:10.1057/palgrave.hep.8300090

Nsamenang, A. B. (2006). Human ontogenesis: An indigenous African view on development and intelligence. *International Journal of Psychology, 41*(4), 293–297. doi:10.1080/00207590544000077

Nyamwaka, E. O., Ondima, P. C., Kemoni, F., & Maangi, E. (2013). The place of music and dance in the reconstruction of African cultural history: A case of the Abagusii of South-Western Kenya. *Research on Humanities and Social Sciences., 3*(10), 113–119.

Nyerere, J. (1982). Education in Tanzania (extracts taken from Education for self-reliance). In B. A. Fafunwa & J. U. Aisiku (Eds.), Education in Africa. A comparative survey (pp. 235–254). London: Allen & Unwin.

Nzelibe, C. O. (1986). The evolution of African management thought. *International Studies of Management & Organization, 16*(2), 6–16. doi:10.1080/00208825.1986.11656427

Nziramasanga, C. T. (1999). Report on the Presidential Commission of Inquiry into Education and Training. Harare: Government Printers.

Odora-Hoppers, C. A., & Makhale-Mahlangu, P. (1998). A Comparative study of the development, integration and protection of indigenous systems in the third world, p. An analytic framework document. Pretoria: Human Sciences Research Council.

Oliphant, A. W. (2000). From Colonialism to Democracy: Writers and Publishing in South Africa. In N. Evans & M. Seeber (Eds.), *The Politics of Publishing in South Africa. Scottsville.* University of Natal Press.

Omona, A.M. (1998). *Sociology of Education.* Department of Distance Education, Makerere: University Kampala.

Onuoha, F. (2015). *Locating African Values in Twenty–First Century Economics.* Available online at: https://www.africanliberty.org/2015/09/09/onuoha-frank-locating-african-values-in-twenty-first-century-economics/

Oommen, T. K. (2001). *Understanding Indian Society: The Relevance of Perspective from Below.* Department of Sociology, University of Pune.

Owusu-Ansah, F. E., & Mji, G. (2013). African indigenous knowledge and research. *African Journal of Disability, 2*(1), 1–5. doi:10.4102/ajod.v2i1.30 PMID:28729984

Oyewumi, O. (1997). The invention of women, p. making an African sense of Western gender discourse. University of Minnesota Press.

Paavola, S., Lipponen, L., & Hakkarainen, K. (2004). Models of innovative knowledge communities and three metaphors of learning. *Review of Educational Research, 74*(4), 557–576. doi:10.3102/00346543074004557

Palumbu, M.V. (2010). Tai Chi for Older Nurses: A Workplace Wellness Pilot Study. *Applied Nursing Research, 25*(1), 54-59.

Paraskeva, J. M. (2016). Curriculum epistemicide, p. Toward an itinerant curriculum theory. New York: Routledge. doi:10.4324/9781315734781

Parker, A., & Van Belle, J. (2017). The iGeneration as Students: Exploring the Relative Access, Use of, and Perceptions of IT in Higher Education. In Liebenberg, J. & Gruner, S. (Eds.), *ICT Education: 46th Annual Conference of the Southern African Computer Lecturers' Association.* Springer.

Patel, N. K., Newstead, A. H., & Ferrer, R. L. (2012). The effects of yoga on physical functioning and health related quality of life in older adults: A systematic review and meta-analysis. *Journal of Alternative and Complementary Medicine (New York, N.Y.), 18*(10), 902–917. doi:10.1089/acm.2011.0473 PMID:22909385

Pauw, B. A. (1963). African Christians and Their Ancestors. In V. E. W. Hayward (Ed.), *African Independent Churches Movements*. Edinburgh House Press.

Pavid, K. (2018). *Rethinking our human origins in Africa, Natural History Museum.* Obtained from: https://www.nhm.ac.uk/discover/news/2018/july/the-way-we-think-about-the-first-modern-humans-in-africa.html

Pearson, S. A. G. (2012). Colonialism and the African experience. In V. B. Khapoya (Ed.), *The African experience* (pp. 99–138). Longman.

Peltier, C. (2018). An application of Two-Eyed Seeing: Indigenous Research Methods With Participatory Action Research. *International Journal of Qualitative Methods, 17*(1), 1–12. doi:10.1177/1609406918812346

Pence, A., & Nsamenang, B. (2008). *A case for early childhood development in subSaharan Africa.* Working Paper No. 51, Bernard van Leer Foundation, The Hague, Netherlands.

Pfeffer, T. (2012). *Virtualization of Universities: Digital Media and the Organization of Higher Education Institutions.* Springer-Verlag. doi:10.1007/978-1-4614-2065-1

Piet, J., & Hougaard, E. (2011). *The Effect of Mindfulness-based Cognitive Therapy for Preventation of Relapse in Recurrent Major Depressive Disorder: A Systematic Review and Meta-analysis.* Clinical.

Pinar, W. F. (2009). *Understanding Curriculum: An Introduction to the Study of Historical and contemporary curriculum discourses* (W. F. Pinar, W. M. Reynolds, P. Slattery, & P. M. Taubman, Eds.). Peter Lang.

Pinar, W. F. (2009). *Understanding Curriculum: An Introduction to the Study of Historical and Contemporary Curriculum Discourses*. Peter Lang.

Pityana, B. (2008). Reflections on 30 years since the death of Steve Biko: A legacy revisited. In C. W. du Toit (Ed.), *The legacy of Stephen Bantu Biko* (pp. 1–14). UNISA.

Pityana, N. (1981). The Black Consciousness Movement and Social Research. In J. Rex (Ed.), *Apartheid and Social Research* (pp. 161–184). UNESCO.

Posnock, R. (1997). How It Feels to Be a Problem: Du Bois, Fanon and the 'Impossible life of the Black Intellectual'. *Critical Inquiry*, *23*(2), 323–349. doi:10.1086/448831

Radebe, X. (2018). New Yoga 4 New Age: Activate Your Personal Energy. Authors House.

Raina, V. K. (1999). Indigenizing teacher education in developing countries: The Indian context. *Prospects*, *29*(1), 5–25. doi:10.1007/BF02736822

Ramachandran, V., & Naorem, T. (2013). What it means to be a Dalit or tribal child in our schools: A synthesis of a six-state qualitative study. *Economic and Political Weekly*, 43–52.

Ramani, S. (2011). Decolonising Knowledge Systems. *Economic and Political Weekly*, 17–19.

Ratcliffe, P. (2001). Race' and Ethnicity Research in Britain: Key Ethical and Political Considerations. In P. Ratcliffe (Ed.), *The politics of social science research: race, ethnicity, and social change*. Palgrave. doi:10.1057/9780230504950_6

Resnik, D. B. (2015). *What is Ethics in Research & Why is it Important?* National Institute of Environmental Health Sciences: https://www.niehs.nih.gov/research/resources/bioethics/whatis/index.cfm

Richards, K., & Govere, E. (2003). Educational legislation in colonial Zimbabwe (1899-1979). *Journal of Educational Administration and History*, *35*(2), 137-151. DOI: . doi:10.1080/0022062032000119822

Rodney, W. (1972). *How Europe underdeveloped Africa*. Bogle-L'Overture Publications.

Roney, R. C. (1996). Storytelling in the Classroom: Some Theoretical Thoughts. *Storytelling World*, *9*, 7–9.

Saigal, A. (2012). Demonstrating a situated learning approach for in-service teacher education in rural India: The Quality Education Programme in Rajasthan. *Teaching and Teacher Education*, *28*(7), 1009–1017. doi:10.1016/j.tate.2012.05.007

Santos, B. S. (2007). Beyond abyssal thinking from global lines to ecology of knowledges. *RE:view*, *30*(1), 1–33.

Sarpong, P. (1991). *Ghana in retrospect: Some aspects of the Ghanaian culture*. African Books Collective Ltd.

Sarpong, P. (2002). *Peoples differ: An approach to inculturation in evangelization*. SubSaharan Publishers.

Sartre, J. P. (1973). Being and nothingness (H. E. Barnes, Trans.). New York: Washington Square Press.

Saunders, P. A., Tractenberg, R. E., & Chertji, R. (2007). Promoting Self-awareness and Reflection through an Experiential Mind-Body Skills Course For First Year Medical Students. *Medical Teacher*, *29*(8), 778–784. doi:10.1080/01421590701509647 PMID:17852720

Saville-Troike, M. (1989). *The ethnography of communication – An introduction*. Basil Blackwell.

Schenck, R., Nel, H., & Louw, H. (2015). *Introduction to participatory community practice*. Unisa Press.

Scully, D., Kremer, J., Meade, M. M., Graham, R., & Dudgeon, K. (1998). Physical exercise and psychological well-being: A critical review. *British Journal of Sports Medicine*, *32*(2), 111–120. doi:10.1136/bjsm.32.2.111 PMID:9631216

Seixas, P. (1993). The community of inquiry as a basis for knowledge and learning: The case of history. *American Educational Research Journal*, *30*(2), 305–324. doi:10.3102/00028312030002305

Senge, P. (1990). *The Fifth Discipline*. Doubleday Currency.

Serrat, O. (2008). *Storytelling*. Available online at: https://www.adb.org/sites/default/files/publication/27637/storytelling.pdf

Shanguhyia, M. S., & Falola, T. (2018). Introduction. In M. S. Shanguhyia & T. Falola (Eds.), The Palgrave handbook of African colonial and postcolonial history. New York: Palgrave MacMillan.

Shizha, E. (2013). Reclaiming our indigenous voices, p. The problem with sub-Saharan African school curriculum. *Journal of Indigenous Social Development*, *2*(1), 1–18.

Silova, I., Millei, Z., & Piattoeva, N. (2017). Interrupting the coloniality of knowledge production in comparative education: Postsocialist and postcolonial dialogues after the cold war. *Comparative Education Review*.

Şimşita, Z. T., Günayb, N. S., & Vayvayc, Ö. (2014). Theory of Constraints: A Literature Review. *Procedia: Social and Behavioral Sciences*, *150*, 930–936. doi:10.1016/j.sbspro.2014.09.104

Singh, S. (2016). Implementing Yoga Education in schools for holistic development of students. *International Journal of Advanced Educational Research*. www.newresearchjournal.com/educational

Sinha, C. (2016). Teaching as a political act: The role of critical pedagogical practices and curriculum. *Human Affairs*, *26*(3), 304–316. doi:10.1515/humaff-2016-0026

Sithole, T. (2016). Frantz Fanon: Africana existentialist philosopher. *African Identities, 14*(2), 177-190. DOI: 5 doi:10.1080/14725843.2015.111738

Slee, P. T. (2002). *Child, Adolescent and Family Development* (2nd ed.). Cambridge University. doi:10.1017/CBO9781139164948

Small, S. (2004). Researching 'mixed-race' experience under slavery: concepts, methods and data. In Researching Race and Racism. Routledge.

Smith, L. T. (1999). Decolonising methodologies: Research and indigenous people. Zed Books.

Smith, L. T. (1999). Decolonizing methodologies, p. Research and indigenous peoples. London: Zed Books University of Otago Press.

Sobel, D. (2004). Place-based education: Connecting classroom and community. *Nature and Listening*, *4*, 1–7.

Sogiba, Z. S. (1996). *Steve Bantu Biko: Politician, 'Historian' and 'Proponent' of African Traditional Religion*. University of Cape Town.

Solórzano, D. G., & Yosso, T. J. (2002). Critical race methodology: Counter-storytelling as an analytical framework for education research. *Qualitative Inquiry*, *8*(1), 23–44. doi:10.1177/107780040200800103

Spitzberg, B. H., & Cupach, W. R. (1988). *Handbook of Interpersonal Competence Research*. Springer-Verlag.

Spivak, G. C. (1990). The post-colonial critic. Viking.

Stals, T. (2001). *The South African Economy: An Evaluation*. http://www.resbank. co.za/address/1996/ad110796html

Stayt, H. A. (1931). *The Bavenda*. Oxford University.

Stones, E. (1966). *An introduction to educational psychology*. Methuen.

Straume, L. V., & Vitterso, J. (2012). Happiness, inspiration and the fully functioning person: Separating hedonic and eudaimonic well-being in the workplace. *Journal of Positive Psychology, 7*, 387-398.

Strydom, F., Kuh, G., & Mentz, M. (2009). Enhancing success in South Africa's higher education: measuring student engagement. *Acta Academica, 42*(1).

Stuart, J. F., Van Niekerk, L. J., McDonald, M. E. W., & De Klerk, D. (1987). *Didactics*. Southern Books.

Styrke, B. (2015). Didactics, dance and teacher knowing in an upper secondary school context. *Research in Dance Education*, *16*(3), 201–212. doi:10.1080/14647893.2015.1036021

Suda, C. (1996). *The Centrality of Women in the Moral Teachings in African Society. Nordic Journal of African Studies*, *5*(2), 71–82.

Sue, D. W., & Sue, D. (1990). *Counseling the Culturally Different: Theory and Practice* (2nd ed.). Wiley.

Surdarshan, P. V., & Balakrishanaiah, S. (2003). *Physical Education Curriculum at Lower Primary Stage- An Evaluation. Research Project*. Regional Institute of Education.

Tanyanyiwa, V. I., & Chikwanha, M. (2011). The role of indigenous knowledge systems in the management of forest resources in Mugabe area, Masvingo, Zimbabwe. *Journal of Sustainable Development in Africa, 3*(3), 132–149.

Tarsons, P. (1958). *Max Weber: The Protestant Ethic and the Spirit of Capitalism*. Charles Scribner's Sons.

Terwel, J. (1999). Constructivism and its implications for curriculum theory and practice. *Journal of Curriculum Studies, 31*(2), 195–199. doi:10.1080/002202799183223

Thaman, K. H. (2003). Decolonizing Pacific Studies: Indigenous Perspectives, Knowledge, and Wisdom in Higher Education. *The Contemporary Pacific, 15*(1), 1–17. doi:10.1353/cp.2003.0032

Thapa, K. (2019). Introduction: Psychological Perspectives on Diversity and Social Development. In J. Pandey, R. Kumar, & K. Thapa (Eds.), *Psychological Perspectives on Diversity and Social Development*. Springer. doi:10.1007/978-981-13-3341-5_1

Thapan, M. (2010). Imagining Citizenship: Being Muslim, Becoming Citizens in Ahmedabad. *Economic and Political Weekly*, 45–50.

The Indigenous World. (2006). *The state has also established an Inter Departmental WorkingGroup on Khoe and San Issues*. Author.

Thiong'o. (1993). *Moving the centre: The struggle for cultural freedoms*. East African Educational Publishers Ltd.

Thobejane, T. D. (2013). *History of Apartheid Education and the Problems of Reconstruction in South Africa Sociology Study*. Academic Press.

Thompson, J.B. (1984). *Studies in the Theory of Ideology, Social and political theory*. Polity Press.

Thompson, L. (1990). *A history of South Africa*. Yale University Press.

Thorat, A., Vanneman, R., Desai, S., & Dubey, A. (2017). Escaping and falling into poverty in India today. *World Development, 93*, 413–426. doi:10.1016/j.worlddev.2017.01.004 PMID:28966435

Tilak, J. B. (2018). Education poverty in India. In Education and development in India (pp. 87-162). Palgrave Macmillan. doi:10.1007/978-981-13-0250-3_3

Tilak, J. B. (2018). The Kothari commission and financing of education. In Education and Development in India (pp. 255-282). Palgrave Macmillan. doi:10.1007/978-981-13-0250-3_8

Torchia, C. (2014, Jan. 27). What next for the rainbow nation? Nelson Mandela's death casts shadow over racially charged South Africans. *The National Post*.

Tordoff, W. (2002). *Government and politics in Africa*. Indiana University Press.

Trevor-Roper, H. (1963). The rise of Christian Europe. The Listener, 70(1809), 871–875.

Trust, S. (2019). Knowledge Valorisation for Inclusive Innovation and Integrated African Development. In S. Oloruntoba & M. Muchie (Eds.), *Innovation, Regional Integration, and Development in Africa. Advances in African Economic, Social and Political Development*. Springer. doi:10.1007/978-3-319-92180-8_7

Tstintziloni, S. (2015). Koula Prastika and her dance school, p. embracing gender, class and the nation in the formative years of contemporary schooling dance education in Greece. *Research in Dance Education, 16*(3), 276–290. doi:10.1080/14647893.2015.1036017

Tuck, E., & Guishard, M. (2013). Un-collapsing ethics: Racialized sciencism, settler coloniality, and an ethical framework of decolonial participatory action research. In T. M. Kress, C. Malott, & B. Porfino (Eds.), *Challenging status quo retrenchment: New directions in critical research* (pp. 3–28). Information Age Publishing.

Tuhiwai-Smith, L. (2012). Decolonizing methodologies, Research and Indigenous People's (2nd ed.). Ottago University Press.

Tukdeo, S. (2019). Towards Reconstruction: Education in Postcolonial India. In *India Goes to School* (pp. 43–56). Springer. doi:10.1007/978-81-322-3957-4_3

Tutu, N. (1989). *The Words of Desmond Tutu*. Newmarket.

Tyack, D., & Cuban, L. (1995). *Tinkering toward utopia: a century of public schools reform*. Harvard University Press.

Uchendu, V. C. (1979). *Education and Politics in Tropical Africa*. St Martin's.

Union of South Africa. (1953). *Bantu Education Act, Act No 47 of 1953*. Retrieved: August, 2017, from http://www.sahistory.org.za/archive/bantu-education-act%2C-act-no-47-of-1953

Unite Nations. (1999). Statement of the UN Committee on Economic, Social and Cultural Rights to the Third Ministerial Conference of the World Trade Organisation (Seattle, 30 November to 3 December 1999).

University of Alabama. (2007). *U.S. Department of Health & Human Services*. https://obssr.od.nih.gov/wp-content/uploads/2016/05/Ethical-Challenges.pdf

Valji, N. (2003). *Creating the nation: The rise of violent xenophobia in the new South Africa* (MA thesis). York University.

Venkatesh, M. (2005). *The effect of yoga on the personality development of students* (Ph.D. thesis). Swami Vivekananda Yoga Anusandhana Samsthana.

Venter, I. S. J., & Van Heerden, S. M. (1992). *The grounding of history of education: an introduction*. Butterworths.

Verhoef, M. (2006). Evaluating a yogic breathing and meditation intervention for individuals living with HIV/AIDS. *American Journal of Health Promotion, 20*(3), 192–195. doi:10.4278/0890-1171-20.3.192 PMID:16422138

Verhoef, M. J., Mulkins, A., & Boon, H. (2005). Integrative health care: How can we determine whether patients benefit? *Journal of Alternative and Complementary Medicine (New York, N.Y.), 11*(supplement 1), s-57–s-65. doi:10.1089/acm.2005.11.s-57 PMID:16332188

Verster, T. L. (1986). *Education themes in time perspective. Part 2*. Butterworths.

Viljoen, R. (2001). Aboriginal Khoikhoi servants and their masters in colonial Swellendam, South Africa, 1745–1795. *Agricultural History, 75*(1), 28–51. doi:10.1525/ah.2001.75.1.28

Vincent, C., & Chikerema, A. F. (2014). Indeginisation of democracy: Harnessing traditional leadership in promoting democratic values in Zimbabwe. *Journal of Power, 2*(1), 67–78.

Vittrup, B. (2016). Early Childhood Teachers' Approaches to Multicultural Education and Perceived Barriers to Disseminating Anti-Bias Messages. *Multicultural Education, 37*(1), 54–67.

Wa Thiong'o, N. (1981). *Decolonizing the mind: The politics of language in African literature*. East African Educational Publishers.

Wane, N. (2008). Mapping the field of Indigenous knowledges in anti-colonial discourse: a transformative journey in education. *Race Ethnicity and Education, 11*(2), 183-197. DOI: . doi:10.1080/13613320600807667

Wane, N. (2011). African indigenous feminist thought: An anti-colonial project. In N. Wane, A. Kempf, & M. Simmons (Eds.), The politics of cultural knowledge (pp. 7-22). Rotterdam: Sense Publishers. doi:10.1007/978-94-6091-481-2_2

Wane, N., & Simmons, M. (2011). Introduction: The politics of cultural knowledge. In N. Wane, A. Kempf, & M. Simmons (Eds.), The politics of cultural knowledge (pp. 1-6). Rotterdam: Sense Publishers.

Wang, C., Bannuru, R., Ramel, J., Kupelnick, B., Scott, T., & Schmid, C. H. (2010). *Tai Chi on Psychology Well-being: systematic review and meta-analysis*. ISCMR.

Wang, C., Collet, J. P., & Lau, J. (2004). The effect of Tai Chi on health outcomes in patients with chronic conditions: A systematic review. *Archives of Internal Medicine, 164*(5), 493–501. doi:10.1001/archinte.164.5.493 PMID:15006825

Warren, D. M. (1991). *Using Indigenous Knowledge in Agricultural Development*. Discussion Paper 127. Washington, DC: World Bank.

Welsing, F. C. (1991). *The ISIS Papers: The keys to the colours*. Third World Press.

West, M. (1976). *Abantu, p. an introduction to the black people of South Africa*. Struik.

WHO. (2009). *Research ethics committee: basic concepts for capacity building*. WHO Press.

Widel, A. (2009). British colonial education in Africa: Policy and practice in the era of trusteeship. *History Compass*, *7*(1), 1–21. doi:10.1111/j.1478-0542.2008.00560.x

Windschitl, M. (2002). Framing constructivism in practice as the negotiation of dilemmas: An analysis of the conceptual, pedagogical, cultural, and political challenges facing teachers. *Review of Educational Research*, *72*(2), 131–175. doi:10.3102/00346543072002131

Wong, P. T. P. (2012). *The human quest for meaning: Theories, research, and applications.* Routledge.

Woolman, D. C. (2001). Educational reconstruction and post-colonial curriculum development. A comparative study of four African countries. *International Education Journal*, *2*(95).

Wrigely, T. (2003). Schools of hope: A new agenda for school improvement. Stoke-on-Trent: Institute of Education Press.

Yeh, G. Y., McCarthy, E. P., Wayne, P. M., Stevenson, L. W., Wood, M. J., Forman, D., Davis, R. S., & Phillips, R. S. (2010). Tai Chi Exercise in Patients with Heart Failure: A Randomized Clinical Trail'. *Archives of Internal Medicine*, *171*(8), 750–757. PMID:21518942

Yosso, T. J. (2005). Whose culture has capital? A critical race theory discussion of community cultural wealth. *Race, Ethnicity and Education*, *8*(1), 69–91. doi:10.1080/1361332052000341006

Young, C. (2012). *The postcolonial state in Africa: Fifty years of independence. 1960–2010.* The University of Wisconsin Press.

Yountae, A. (2017). The decolonial abyss: Mysticism and cosmopolitics from the ruins. Fordham University Press.

Zagzebski, L. (2004). Foreknowledge and Free Will. *Stanford Encyclopedia of Philosophy.* Retrieved: June, 2012, from: https://plato.stanford.edu/entries/free-willforeknowledge

Zimmerman, L. J. (2005). First be humble: working with indigenous people and other descendant communicates. In Indigenous Archaeologies: Decolonising Theory and Practice. Academic Press.

Zondi, N., & Ntshangase, M. (2013). Shifting Boundaries in Respect of Perceptions Governing Women and the Burning of *Impepho. Alternation (Durban)*, *20*(2), 213–235.

Zulu, I. M. (1993). *The Ancient 'emetic Roots of Library and Information Science. California Univ.* Afro-American Studies Center. Nov.

Zuze, T. L. (2012). The Challenge of Youth-To-Work Transitions: An International Perspective. In H. Perold, N. Cloete & J. Papier (Eds.), Shaping the Future of South Africa's Youth: Rethinking Post-School Education and Skills Training (pp. 42–59). Cape Town: Centre for Higher Education Transformation (CHET).

About the Contributors

Ntokozo Mthembu lectures Sociology at the University of South Africa. His research interests include social capital, survival theory, youth perspectives, and indigenous knowledge systems.

* * *

Thizwilondi Daswa is a Lecturer in the Department of Indigenous Knowledge Systems, School of Human and Social Sciences, University of Venda, Limpopo Province, South Africa Academic Qualifications: Masters of Arts in African Studies (UNIVEN), BED Hons, Education Training and Development(UJ), Further diploma in Education (RAU), Secondary Teachers Diploma (VECO), Diploma in Theology (Zoe Logos Institute) Employment History: Worked for the department of Education as an educator for 23yrs, currently an IKS lecturer at the University of Venda. Academic Experience: Teaching and supervising undergraduate and postgraduate students. I have supervised honours dissertation (BIKS) in the University of Venda.

Jairos Gonye holds a PhD In English (University of Venda, South Africa). He is currently an Associate Professor at Great Zimbabwe University and also a Research Associate with the University of the Free State, South Africa. His research interests include postcolonial literature, dance representations and pedagogy. He has published both locally and internationally in these related areas.

Stewart Lee Kugara is a senior lecturer at the University of Venda in the Centre for African Studies and a former lecturer in the School of Law. He holds a PhD in African Studies, a Master's Degree in Law (LLM) and LLB. He is an interdisciplinary, intradisciplinary and multidisciplinary author with academic publications straddling the following areas; intellectual property, traditional leadership, African traditional health and beliefs, knowledge management, African conflicts management, cultural astronomy, human rights, customary law, poverty and development studies, African philosophy and cultural philosophy.

Justice Makhanikhe is a lecturer at the University of Venda in the Department of African Studies. He holds a Master of Arts degree in African Studies.

Tsetselelani Mdhluli is an NGAP Indigenous Knowledge Systems (IKS) lecturer in the Department of Cultural and Political Studies at the University of Limpopo and is a candidate for Doctor of Philosophy in African studies. She holds a Master's Degree in African Studies, (Hons) Degree in Gender Studies and degree in International Relations. She has been working with the University of Venda as a research assistant under the Community Engagement Directorate and as a mentor for students studying Indigenous Knowledge in the Centre for African Studies. She is an author in the following areas: poverty and development studies, African philosophy, cultural philosophy, gender, humanitarianism, and civil society organisations, indigenous knowledge systems and knowledge management.

Nathan Moyo completed his PhD in Curriculum Studies with the University of Johannesburg, South Africa. He is a Senior Lecturer in Curriculum Studies at Great Zimbabwe University, Zimbabwe. He teaches Curriculum Theory and History Education. His research interests include curriculum theory, history education, postcolonial theory, and social justice education. His most recent publications include book chapters in the The Palgrave Handbook of Race and the Arts in Education (2018) and The Palgrave Handbook of History and Social Studies Education (forthcoming). Nathan has published in leading international journals such as Journal of Education Policy, Research Papers in Education, Research in Dance Education and Pedagogy, Culture and Society.

Faith Musvipwa is PhD candidate for Sociology at the University of Venda.

Zingisa Nkosinkulu is an artist, art historian, and curator whose areas of study and practice include contemporary African art, black aesthetics, decolonial aesthetics/aestheSis, Hip Hop and Jazz, graffiti/street art, decolonial epistemic perspective, Xhosa aesthetics, and black existentialism. He is especially concerned with the depiction of the black lived experience and its social formation in black creative expression practices, in particular, Nkosinkulu is interested in the relation between decolonial thought and experimental art, and has been preoccupied with fusing these fields to illuminate the power of their synergy. He addresses these concerns by way of art-making, art exhibitions and criticism with special focus on Mary Sibande's visual artworks, Fanonian discourse, Jean Michel Basquiat, Tupac Shakur, and Édouard Glissant. Nkosinkulu was born in Eastern Cape, South Africa, he is currently a lecturer of Art History in the Department of Art at UNISA School of Arts where he is also enrolled for his Ph.D. in Art. Nkosinkulu believes he has a calling to use art and decolonial thought as a mediator to transcend beyond the living conditions and colonial legacies in South African and the rest of the world. As the founder of Art Curate Afrika (ACA), with experience in Visual Art and Project Management since 2007, he has held three solo exhibitions of which he self-curated at the Grahamstown Arts Festival. Nkosinkulu's mission is to use decolonial art, Xhosa tradition and writing to bring new hope, share creative ideas and encourage people to be artistic, educated, and conscious individuals.

Jah-Xolani Radebe is currently masters candidate and published Author for *New Yoga 4 New Age.*

Ndidzulafhi Ramavhunga is a senior lecturer at Venda in the Centre for African Studies. She is also a teacher by profession for 31 years. She holds a PhD in African Studies, a Master's Degree MED from NICE (Australia), B.A HONS from UNIVEN, B.A, from Univen and Further diploma in Educational guidance RAU, She is an author of Reflections on practices of U Laya Nwana: Towards Afrosensed Approach. She is supervising undergraduate and postgraduate students in the University of Venda.

Xolani Mathews Shange is a PhD candidate in development studies at the University of South Africa. He has accumulated experience in South Africa's developmental landscape associated with implementation of urban and rural development programmes. He also participated in social movements, teaching and lecturing. His research interests include Indigenous Knowledge Systems and Social Policy with particular focus on local governance, urban planning, skills development, curriculum development and food security.

Rendani Tshifhumulo is a senior lecturer and a coordinator of Sociology section at the university of Venda. My area of research includes contemporary social problems, culture and health. I have taught medical sociology for more than 15 years.

Livhuwani Tshikukuvhe is a lecturer of indigenous knowledge systems(IKS) at the University of Venda.

Index

U

V

W

Y

www.ingramcontent.com/pod-product-compliance
Lightning Source LLC
LaVergne TN
LVHW081328300926
849161LV00008B/1909